# Synthesis Lectures on Human Language Technologies

**Series Editor**

Graeme Hirst, Department of Computer Science, University of Toronto, Toronto, Canada

The series publishes topics relating to natural language processing, computational linguistics, information retrieval, and spoken language understanding. Emphasis is on important new techniques, on new applications, and on topics that combine two or more HLT subfields.

Punyajoy Saha · Mithun Das ·
Animesh Mukherjee

# Online Hate Speech

Analysis, Detection, and Mitigation

Punyajoy Saha
Department of Computer Science
and Engineering
Indian Institute of Technology Kharagpur
Kharagpur, West Bengal, India

Mithun Das
Department of Computer Science
and Engineering
Indian Institute of Technology Kharagpur
Kharagpur, West Bengal, India

Animesh Mukherjee
Department of Computer Science
and Engineering
Indian Institute of Technology Kharagpur
Kharagpur, West Bengal, India

ISSN 1947-4040　　　　　　ISSN 1947-4059　(electronic)
Synthesis Lectures on Human Language Technologies
ISBN 978-3-031-86594-7　　ISBN 978-3-031-86595-4　(eBook)
https://doi.org/10.1007/978-3-031-86595-4

© The Editor(s) (if applicable) and The Author(s), under exclusive license to Springer
Nature Switzerland AG 2026

This work is subject to copyright. All rights are solely and exclusively licensed by the Publisher, whether the whole or part of the material is concerned, specifically the rights of translation, reprinting, reuse of illustrations, recitation, broadcasting, reproduction on microfilms or in any other physical way, and transmission or information storage and retrieval, electronic adaptation, computer software, or by similar or dissimilar methodology now known or hereafter developed.
The use of general descriptive names, registered names, trademarks, service marks, etc. in this publication does not imply, even in the absence of a specific statement, that such names are exempt from the relevant protective laws and regulations and therefore free for general use.
The publisher, the authors and the editors are safe to assume that the advice and information in this book are believed to be true and accurate at the date of publication. Neither the publisher nor the authors or the editors give a warranty, expressed or implied, with respect to the material contained herein or for any errors or omissions that may have been made. The publisher remains neutral with regard to jurisdictional claims in published maps and institutional affiliations.

This Springer imprint is published by the registered company Springer Nature Switzerland AG
The registered company address is: Gewerbestrasse 11, 6330 Cham, Switzerland

If disposing of this product, please recycle the paper.

# Preface

Social media platforms like Twitter, Gab, and Facebook have become ubiquitous in our social lives. These platforms allow users to instantly share their ideas and opinions at practically no cost. Malicious actors have already used this opportunity to cause online harm in various ways. These platforms often link to scenes like the Rohingya genocide in Myanmar, Islamophobic mob violence in Sri Lanka, and the Pittsburgh synagogue shooting. For quite some time now, hate speech has been considered as one of the significant issues poisoning the online social media environment. For a proper and healthy functioning of these platforms there is a dire need to have computational techniques to understand how hate speech spreads across them. Further, such hateful content should not go undetected and algorithmic methods need to be developed to sieve through billions of posts to auto-discard such content. Third, there is a need to know how hateful users behave and what could be an effective way to mitigate hate speech. This book looks at the recent advances and issues surrounding the analysis, spread, detection plus the challenges therein and mitigation of hate speech in online social media.

We begin with an introductory chapter discussing the definitions of hate speech, the worldwide rise in online hate content, the position taken by peace-mediating organizations, like the UN, the different categories of hate speech and the varied media in which they are found. This chapter also describes the various contexts of hate speech and finally sets the expectations of the readers as to what they could know/learn from this book.

In the next chapter we shall demonstrate the prevalence of hate speech on both mainstream platforms like X, Facebook, etc. as well as fringe platforms like Gab, 4Chan etc. We examine this from different perspectives. First, we analyze the language used in hate speech and its evolution over time. Next, we focus on the users posting such content and their network characteristics. We also discuss about other forms of harmful content that are more nuanced and difficult to trap like fear speech and dangerous speech. Finally we show how certain newly emerging platforms have become the breeding grounds of hate.

In the subsequent chapter, we plan to discuss the kind of content that needs attention on social media platforms and explore methods that detect such hateful content automatically, ranging from traditional machine learning models (e.g., logistic regression, SVM, etc.) to deep learning-based models (LSTM, BERT, etc.). Many of these machine learning architectures are quite complex. We therefore plan to write this section so that is easily accessible to beginners.

Next we shall discuss about the challenges in hate speech detection. We shall point out the bias and fairness issues related to the datasets used and the models employed. We shall also present a discussion on explainability of these detection models.

In the last chapter, we shall look into the mechanisms for mitigating the problem of hate speech. These include straightforward approaches, deletion of posts and accounts, shadow banning to softer methods like counterspeech, i.e., responses countering hate speech. For each mitigation strategy we plan to demonstrate its primary mechanism, advantages, and drawbacks. In the section on counterspeech, we shall also focus on how generative AI can be applied to produce automatic counterspeech. We shall also discuss how some of these models can be improved using data augmentation and controller techniques. Finally, we shall end with a discussion about how LLMs can be used to generate counterspeech.

**A note of warning to readers**: This book contains examples of the hate speech that we aim to detect and mitigate, and readers will undoubtedly find them offensive and possibly triggering. We include them nonetheless, as they are necessary to demonstrate the nature of the problem and the methods that we use to counter them.

Kharagpur, India  
December 2023

Animesh Mukherjee  
Mithun Das  
Punyajoy Saha

# Acknowledgements

A large number of people have explicitly or implicitly helped in the development of this book. While it is impossible to articulate the support extended by each of them, it is absolutely necessary to mention some without whom this book would have perhaps never been written.

The credit for identifying the necessity for studying online hate speech as a research problem in our group goes to **Binny Mathew**. Almost six years back, while the research community was still unsure about the potential of this research, Binny was firm and insisted to pursue a Ph.D. this area. Soon we felt that the problem was exceptionally intriguing and computational methods were still scarce. In fact, many of the chapters in this book will have references to the works that Binny did in his Ph.D. tenure. All the three authors would like to express their gratitude to Binny for his relentless contributions to the field as well as for introducing this highly potent area of research into the group.

While we were working on this area we felt the need of disseminating the findings with a larger audience given the societal importance and public relevance of the topic. We were fortunate to have been able to successively deliver tutorials on this topic at three top conferences—ICWSM 2021, AAAI 2022, and WSDM 2023. Preparing for these tutorials helped in enriching our knowledge about the literature that complimented our own works. Moreover interactions with the audience during these tutorials allowed us to view the problem from different perspectives many of which have been useful in drafting this book. We are grateful to the organizers of these conferences for giving us this platform.

Throughout our journey in this research we were very fortunate to have been able to collaborate with top notch scientists across the globe. Some of them whom we are especially thankful to are—(Pawan Goyal, IIT Kharagpur), (Chris Biemann, University of Hamburg), (Kiran Garimella, University of Rutgers), (Soumya Sarkar, Microsoft-IDC), (Manish Gupta, Microsoft-IDC), and (Vikram Gupta, Sharechat).

We are also grateful to all colleagues whose research works we have referred to in this book to compliment our own studies. We believe that they have all been very instrumental in fighting this malice that has overwhelmingly inflicted the different social networks in recent times.

Animesh Mukherjee would like to express his heartfelt gratitude to his wife Soumita and his daughter Anugyaa who have been a continuous source of inspiration and support enabling him to cope up all the hurdles of disability and seamlessly pursue research. He would also like to thank his parents and teachers who have inculcated the right moral values in him thus empowering him to choose this sensitive area of research.

Mithun Das would like to convey his sincere appreciation to his supervisor, Prof. Animesh Mukherjee, whose guidance has been pivotal in shaping his research acumen. He owes a debt of gratitude to his wife, Keya Das, for her unwavering encouragement and steadfast support. Gratitude fills Mithun's heart as he expresses his most profound thanks to his precious son, Kingshuk, for gracefully entering his life. Lastly, he acknowledges the profound influence of his father, mother, brother, and sister on both his life and achievements. They have played a crucial role in shaping his character, and any success he has achieved is a testament to their profound impact.

Punyajoy extends his sincere thanks to Prof. Animesh Mukherjee, acknowledging the pivotal role the supervisor has played in shaping his research proficiency. Furthermore, Punyajoy expresses deep appreciation for the profound influence of his father, mother, brother, and sister on both his life and achievements. Their instrumental role in shaping his character is undeniable, and the success he has garnered is a testament to their profound impact.

# Contents

**1 Introduction** .................................................................... 1
   1.1 Hate Speech: What Is It? ................................................ 1
      1.1.1 Indicators of Hate Speech ....................................... 2
      1.1.2 Directed Versus Generalized Hate ............................. 3
      1.1.3 Explicit Versus Implicit Hate ................................... 3
   1.2 The Rise of Hate Content in the Online World ....................... 4
   1.3 The UN Mandate ........................................................ 5
      1.3.1 Monitor Hate Speech ............................................ 5
      1.3.2 Mitigate Hate Speech ............................................ 6
      1.3.3 Manage Outcomes ................................................ 6
   1.4 Hate Speech Categorization ............................................. 8
   1.5 Medium of Hate Speech ................................................. 8
   1.6 Contexts of Hate Speech ................................................ 9
      1.6.1 Boundaries of Hate Speech and Intersection with Free Speech ... 9
      1.6.2 Focused Groups .................................................. 10
      1.6.3 Emerging Communities ......................................... 11
   1.7 What to Expect from the Rest of This Book .......................... 11
   References ................................................................... 12

**2 Analysis** ........................................................................ 15
   2.1 Spread of Hate Speech Across Online Platforms ..................... 15
      2.1.1 Moderated Platforms ............................................ 15
      2.1.2 Platforms with Lax Moderation ................................ 17
      2.1.3 Other Forms of Harmful Content .............................. 20
      2.1.4 Newly Emerging Platforms ..................................... 22
   References ................................................................... 23

# 3 Detection ... 25
## 3.1 Keyword-Based Techniques ... 27
## 3.2 Machine Learning Techniques ... 28
### 3.2.1 Extracting Meaningful Features ... 28
### 3.2.2 Popular Algorithms ... 29
### 3.2.3 Limitations ... 30
## 3.3 Deep Learning Techniques ... 30
## 3.4 Transformer-Based Techniques ... 31
## 3.5 Hybrid Techniques ... 32
### 3.5.1 HurtBERT ... 32
### 3.5.2 HateBERT ... 34
### 3.5.3 HateNet ... 35
## 3.6 Evaluation of Hate Speech Detection Systems ... 35
### 3.6.1 Evaluation Metrics ... 38
### 3.6.2 HateCheck Test Suite ... 41
## 3.7 A Review of Key Results ... 42
### 3.7.1 Results from Machine and Deep Learning Models ... 42
### 3.7.2 Performance of HurtBERT ... 45
### 3.7.3 Performance of HateNet ... 45
### 3.7.4 Performance of HateBERT ... 45
## 3.8 Multimodal Information ... 47
### 3.8.1 Image-Based Hate Speech Detection ... 47
### 3.8.2 Audio-Based Hate Speech Detection ... 48
### 3.8.3 Hate Meme Detection ... 48
### 3.8.4 Hate Video Detection ... 49
## 3.9 LLMs for Hate Speech Detection ... 49
### 3.9.1 Prompt Design ... 51
### 3.9.2 Examples in Action ... 51
## References ... 52

# 4 Challenges in Hate Speech Identification ... 57
## 4.1 Pitfalls of Model Evaluation ... 57
### 4.1.1 Caveats of High Performance Values ... 57
### 4.1.2 Impact of User Distribution on Model Generalization ... 60
## 4.2 Bias and Fairness ... 61
### 4.2.1 Bias ... 61
### 4.2.2 Fairness ... 62
### 4.2.3 Fairness Versus Bias ... 63
### 4.2.4 Measuring Bias in Hate Speech Detection Models ... 64
### 4.2.5 How to Make Hate Speech Detection Models More Fair and Less Biased ... 65

| | 4.3 | Explainability | 66 |
|---|---|---|---|
| | | 4.3.1 Types of Explainability | 67 |
| | | 4.3.2 Common Techniques for Explainability | 69 |
| | | 4.3.3 Evaluating Rationales | 72 |
| | | 4.3.4 Explainable Hate Speech Detection | 73 |
| | References | | 77 |
| 5 | **Mitigation** | | 81 |
| | 5.1 | Banning or Suspension | 82 |
| | | 5.1.1 Efficacy of Reddit Ban | 82 |
| | 5.2 | Counterspeech | 86 |
| | 5.3 | Generation of Counterspeech | 88 |
| | | 5.3.1 Generation Models | 89 |
| | | 5.3.2 Decoding Methods | 90 |
| | | 5.3.3 Evaluation Metrics | 91 |
| | | 5.3.4 Human Evaluation | 93 |
| | 5.4 | Comparison Across Generation Strategies | 94 |
| | | 5.4.1 Best Model | 95 |
| | | 5.4.2 Best Decoding Method | 97 |
| | | 5.4.3 Best Model-Decoding Combination | 97 |
| | 5.5 | Generate, Prune and Select | 99 |
| | | 5.5.1 Proposed Model | 99 |
| | | 5.5.2 Experiments | 102 |
| | | 5.5.3 Results | 103 |
| | 5.6 | CounterGeDi: A Controllable Approach to Generate Counterspeech | 104 |
| | | 5.6.1 Proposed Model | 104 |
| | | 5.6.2 Experiments | 107 |
| | | 5.6.3 Results | 110 |
| | 5.7 | Knowledge-Grounded Counter Speech Generation | 113 |
| | | 5.7.1 Proposed Model | 115 |
| | | 5.7.2 Experiments | 117 |
| | | 5.7.3 Results | 118 |
| | 5.8 | LLMs for Mitigation | 119 |
| | | 5.8.1 Models Used | 120 |
| | | 5.8.2 Prompting Strategies | 121 |
| | | 5.8.3 Additional Datasets | 122 |
| | | 5.8.4 Evaluation Metrics | 123 |
| | | 5.8.5 Results | 125 |
| | References | | 131 |
| **Appendix A: Data Repository** | | | 139 |

# Introduction 1

## 1.1 Hate Speech: What Is It?

There is an abundance of the definition of the term 'hate speech' in the literature. The United Nations defines it as follows.[1]

**Definition 1.1** Any kind of communication in speech, writing or behaviour, that attacks or uses pejorative or discriminatory language concerning a person or a group based on who they are, in other words, based on their religion, ethnicity, nationality, race, colour, descent, gender or other identity factor.

Wikipedia adopts the definition of hate speech from the Cambridge dictionary which is as follows.[2]

**Definition 1.2** Public speech that expresses hate or encourages violence towards a person or group based on something such as race, religion, sex, or sexual orientation.

According to the committee of ministers of the council of Europe, hate speech is understood as follows.[3]

**Definition 1.3** All types of expression that incite, promote, spread or justify violence, hatred or discrimination against a person or group of persons, or that denigrates them, because of their real or attributed personal characteristics or status such as race, colour, language,

---

[1] https://shorturl.at/aegmT.
[2] https://en.wikipedia.org/wiki/Hate_speech.
[3] https://shorturl.at/mpsv1.

religion, nationality, national or ethnic origin, age, disability, sex, gender identity and sexual orientation.

Together the tech giants, X, Meta, Microsoft and Google follow a very similar definition. For instance, Meta defines hate speech as follow.[4]

**Definition 1.4** Anything that directly attacks people based on what are known as their "protected characteristics"—race, ethnicity, national origin, religious affiliation, sexual orientation, sex, gender, gender identity, or serious disability or disease.

Various other dictionaries such as the Merriam Webster dictionary, the MacMillan dictionary and the Oxford Learner's dictionary present similar definitions of hate speech which do not vary much from the above ones taken together.

The NLP literature also discusses an array of definitions many of which are often synthesised from a mixture of the above. For instance the authors in Waseem et al. [1] points out that there are two key questions determining the definition of hate/abusive speech. These are 'is the language directed towards a specific individual or entity or is it directed towards a generalized group?' and 'is the abusive content explicit or implicit?' The authors in Rottger et al. [2] introduce two paradigms of hate speech—*descriptive* and *perspective*. The descriptive paradigm stresses on the subjectivity of the hate content whereas the perspective paradigm focuses on one specific belief encoded in the hate speech.

Overall, while there are a variety of definitions for hate speech, an underlying common theme binds them together. The first factor is that hate content expresses hatred (either explicit or implicit) against an individual or a group. The second factor is that hate content often incites violence and promotes outgroup prejudice. Finally, the individuals/groups targeted are often characterized by protected attributes like race, sex, religion, age, nationality, physical disability etc.

### 1.1.1 Indicators of Hate Speech

From the definitions listed above, it becomes apparent that there are certain indicators that contribute to the different components of hate speech. Broadly, these are, as Fortuna and Nunes [3], suggest—hate speech is most often hurled toward specific targets/victims, (2) the objective of hate speech is to instigate violence (3) hate speech is also meant to establish superiority by attacking the minority or marginalised groups, and (4) hate speech is often camouflaged by sarcasm and humour. A hate speech post can have either one or multiple of these components present in it. The presence of multiple components often makes the post complex and therefore hard to algorithmically detect.

---

[4] https://about.fb.com/news/2017/06/hard-questions-hate-speech/.

1.1 Hate Speech: What Is It?

In particular, the indicators should be relevant to the phenomenon of hate speech, it should be interpretable and should remain a valid point of comparison over time [4]. Broadly there are two schemes of attacks—first, based on the sex, gender or sexual preference of a person trying to humiliate, objectify or destroy the victim's reputation often making them vulnerable, and second, based on the marginalization of groups including racial offence, Islamophobia, xenophobia, antisemitism, sinophobia plus discrimination toward people with ethnicity, nationality, disability or disease. The discrimination could be of various types ranging from denial of human rights, dehumanization, spreading of manipulative and false information, associating negative stereotypes, promoting outgroup prejudice etc.

### 1.1.2 Directed Versus Generalized Hate

Depending on the number of victims/targets hate speech can also be classified into two types [1]. These are as follows.
*Directed hate*: In this case, hate speech is expressed toward a specific individual or entity. Example "@usr4 your a fucking queer faggot bitch".
*Generalized hate*: In this case the hate speech is expressed toward a general group of individuals who share a common protected characteristic, e.g., ethnicity or sexual orientation. Example: "—was born a racist and—will die a racist!—will not rest until every worthless nigger is rounded up and hung, niggers are the scum of the earth!! wPww WHITE America".

### 1.1.3 Explicit Versus Implicit Hate

Another important type of categorization of hate content is based on whether they are explicit or implicit [1]. In linguistics, the former is analogous to the denotation or the literal meaning of a word while the latter corresponds to connotation, i.e., the socio-cultural association of the words.
*Explicit hate*: Explicit hate speech can be either directed or generalized; however it is often indicated by a specific set of keywords (e.g., swear words, racial slurs etc.). Examples include "Go kill yourself", "You"re a sad little fuck".
*Implicit hate*: Implicit hate speech can again be either directed or generalized; however, no indicative keywords may be present making the identification of hate speech difficult. Most often, the content has elements of sarcasm, humour or metonymy. Example: "Gas the skypes".

## 1.2 The Rise of Hate Content in the Online World

Online fora have played an instrumental role in connecting people across geographies to share and discuss opinions. Social media is a digital technology that allows the sharing of information and ideas through text and visuals on virtual networks and communities. However, of late, social media has been increasingly used by bad actors to spread hate and inflammatory speech. Such content is linked with the global increase in violence[5] including mob lynchings, mass shootings, riots and ethnic cleansing. Some of the glaring examples include (but not limited to) (a) the Pittsburgh synagogue shooting[6] which was committed by Robert Bowers, an active user of the Gab social network that has a lax moderation attracting extremists banned from other platforms, (b) the Rohingya genocide in Myanmar[7] which was largely propelled by the military leaders and Buddhist nationalists through a widespread campaign of ethnic cleaning on Facebook, (c) the anti-refugee Facebook posts by the far-right Alternative for Germany party which was found to be correlated to attacks on refugees [5], (d) the Islamophobic riots in Sri Lanka[8] which have been largely attributed to the hateful posts on Facebook and WhatsApp, and (e) the spates of mob lynchings in India[9] which have been shown to be connected to rumors spread on WhatsApp.

Social media platforms often allow conspiracy theorists and fringe websites to reach an audience far wider than their core readership. Since these platforms garner most of their revenues from advertisements, it is in their direct interest to allow advertisers to target audiences who would spend time and consume the content on these ad websites. For instance, in February 2023, as many as 10 hateful ads against the LGBTQ+ community in Ireland submitted to Facebook, TikTok and YouTube were approved for publication by these platforms.[10] All the three platforms accepted the "burn all gays" ad and other similar ads upholding violence against transgender women. Millions of dollars have been spent by political parties to run ads on social media platforms prior to an election.[11] Many of these ads were hateful and untrue.[12] However, the platforms did not take any action to prevent the dissemination of

---

[5] https://www.washingtonpost.com/nation/2018/11/30/how-online-hate-speech-is-fueling-real-life-violence/.

[6] https://www.nytimes.com/2018/10/28/us/gab-robert-bowers-pittsburgh-synagogue-shootings.html.

[7] https://www.fortifyrights.org/downloads/Fortify_Rights_Long_Swords_July_2018.pdf.

[8] https://www.reuters.com/article/us-sri-lanka-clashes-socialmedia/sri-lanka-to-lift-social-media-ban-minister-idUSKCN1GP2LO.

[9] https://www.washingtonpost.com/graphics/2018/world/reports-of-hate-crime-cases-have-spiked-in-india/?utm\_term=.2712573765d9.

[10] https://www.irishtimes.com/technology/big-tech/2023/02/23/hateful-ads-submitted-to-facebook-tiktok-and-youtube-approved-by-platforms/.

[11] https://www.newyorker.com/tech/annals-of-technology/the-problem-of-political-advertising-on-social-media.

[12] https://www.theguardian.com/world/article/2024/may/20/revealed-meta-approved-political-ads-in-india-that-incited-violence.

such content arguing that such a ban would curtail the freedom of speech. In fact, such dissemination of hateful and fake content is routinely promoted by the algorithms especially if they are found to steer the ad revenue ad revenue on these platforms. Such completely unfettered speech presents a barrier to participation for the minorities that are targeted.

## 1.3 The UN Mandate

In 2019, the UN Secretary General António Guterres observed that "hate speech is a menace to democratic values, social stability and peace." He appointed his Special Adviser on the Prevention of Genocide to coordinate this plan in collaboration with the UN Working Group on Hate Speech comprising representatives from 16 countries. In addition, the committee also includes UN Country Teams and the peace operations and political missions teams across the globe. Together the committee has come up with the following strategy and plan of action laying out 13 commitments. We organise these commitments into a monitor-mitigate-manage workflow as follows.

### 1.3.1 Monitor Hate Speech

- **COMMITMENT 1**: Monitoring and analysing hate speech. Three actions were suggested in response to this commitment. The first among these is to elevate the knowledge of the UN staff so that they have a full understanding of the objectives and goals for setting up this Strategy, the definition and severity of hate speech and the legal repercussions and actions that can be taken by the member states as per the international human rights law. The second action suggests undertaking a massive data collection drive including both offline and online data representing the context of hate at the country level in the form of an account of historical, political and socioeconomic grievances, and intergroup tensions and violence. The third action suggests the quantitative as well as qualitative analysis of the collected data to understand the hate speech trends at country, region and local contexts. As a part of the analysis identify the gender of the instigators, targets/victims, audiences and challengers to make an assessment of the overall impact of the hate speech.
- **COMMITMENT 2**: Addressing root causes, drivers and actors of hate speech. The first action suggested here is to identify the lead instigators and spreaders of hate speech in a country or a region. The next task is to dig out the root causes that motivated them to become bad actors. Such causes often include economic inequality in the country, absence of basic amenities and civic space, online disinhibition or other historical/current grievances that need to be addressed at the grass root level by the country's leadership. The second action point suggested is to sanction projects and programmes to drive away the root causes of hate speech and promote a culture supporting mutual respect, social cohesion, inclusion and unbiased discussion of controversial opinions.

- **COMMITMENT 6**: Using technology. The UN would encourage extensive research in online hate speech dissemination and the study of the factors that drive individuals/groups toward violence. It would also train its own staffs involved in data handling to understand the global trends in the spread of hate speech. Another important commitment is to promptly inform the giant social media platforms or those that are very popular in a country about the hateful events that might be an immediate cause of violence.
- **COMMITMENT 12**: Building the skills of United Nations staff. The UN would strive its best to train its key staff in dealing with all forms of hate speech and hate crime all across the world. Compulsory capacity-building programmes would be hosted to train these staffs so that they are able to identify (i) the root causes and drivers of hate speech, (ii) the immediate and future trends of a hate event, and (iii) the impact of countermeasures and the extent of their success.

### 1.3.2 Mitigate Hate Speech

- **COMMITMENT 4**: Convening relevant actors. The UN is committed to act as the convener for countering hate speech at the country level. It would find the best channels that can promote counterspeech which can include media, political persons, parliamentarians, celebrities, private sector representatives etc. It should also strive to reframe the hate speech policies in a country so that they are in concordance with international human rights laws.
- **COMMITMENT 7**: Using education as a tool for addressing and countering hate speech. From time to time, the UN would be committed to conduct teacher and student training programmes to spread awareness about human rights in civil society. UN is also committed to using meaningful technology and education to empower women, girls and youths who are often vulnerable to hate crimes. Another active area would be to steer the elimination of biased study materials especially from history education.

### 1.3.3 Manage Outcomes

- **COMMITMENT 3**: Engaging and supporting the victims of hate speech. The UN is committed to expressing solidarity to individuals or groups who have been victims of hate speech especially those who faced discrimination and violence. It also includes human rights-centric measures to counter the negative effects of hate violence and empowerment of the victim group. Wherever appropriate, the UN officials would help the victims to file necessary litigation against the discrimination and hate violence faced by them. The UN also regularly scrutinizes national laws related to hate speech and how these interact

## 1.3 The UN Mandate

with the freedom of speech of the country. If required they advocate and suggest reforms in the law to protect the victims of hate speech.

- **COMMITMENT 5**: Engaging with new and traditional media. Wherever possible the UN will tie up with the mainstream media like newspapers, television etc. while also opening up new connections with other and possibly more popular forms like online social media. The UN would encourage plurality in journalism and push for journalistic freedom and editorial independence. The whistle-blowers reporting about hate speech violence should be extended all necessary protection. Finally, UN is committed to promoting only ethical and responsible journalism whereby the reporters and editors should be fully accountable for what the report is.

- **COMMITMENT 8**: Fostering peaceful, inclusive and just societies to address the root causes and drivers of hate speech. UN would strive to foster peacekeeping campaigns in the form of media interviews, social media posts, nationwide anti-hate posters and placards, targeted programmes on radio and television, student outreach programmes and motivational talks from opinion leaders. UN would also promote intergroup dialogue, cultural exchange programmes and joint activities to enhance cultural cohesion. There should be a dedicated mission to use computer algorithms to automatically generate counterspeech against hateful posts to reduce misconceptions among groups in the online world. Recent developments in large language models like ChatGPT can be used for this purpose.

- **COMMITMENT 9**: Engaging in advocacy. The UN should establish and expand the policies to address future incidents of hate speech. The UN should advocate all measures to refrain all its actors from disseminating hate speech, hostility or violence. Similarly, it should also advocate the expression of solidarity with the victims of hate speech and take adequate measures to restore their dignity.

- **COMMITMENT 10**: Developing guidance for external communications. In the event of a severe hate speech episode, the UN would be committed to developing a targeted communications plan to address, counter and mitigate the impact of the hate speech through a peacekeeping operation or a high-profile political mission. The UN would regularly investigate the composition of audiences exposed to hate speech and counterspeech and have automatic techniques in place to measure the contextual factors influencing their receptiveness of such messages and whether the counterspeech is able to dilute the effect of the harm brought by hate speech.

- **COMMITMENT 11**: Leveraging partnerships. The UN would create and leverage partnerships with government organizations, NGOs, regional and multilateral organizations, religious bodies and most importantly tech industries to fulfil the mission of complete eradication of hate speech and hate violence.

- **COMMITMENT 13**: Supporting Member States. Whenever requested, the UN would extend all necessary technical support and assistance to any Member State in matters related to legislation, policy formation/revision and mechanisms to tackle and counter hate speech and violence. UN would also launch dedicated missions for capacity-building of key State actors like judges, prosecutors, defense lawyers and law enforcement officers on international human rights norms and guidelines relating to hate speech, especially the type of violence that amounts to a criminal offence.

This three-step workflow can be a first step in guiding how hate speech should be handled at both the local and the global scale.

## 1.4 Hate Speech Categorization

One of the popular categorizations of online hate content has been presented by the authors in Fortuna and Nunes [3]. We shall outline the key categories summarised from their work. Since we have already defined hate speech in the beginning of this chapter, here we shall primarily discuss about the other close variants.

- *Abusive speech*: Abusive speech is an umbrella term that is used to refer to any hurtful language like hate speech, derogatory language and profanity.
- *Toxic speech*: Toxic speech usually corresponds to posts that are disrespectful, rude or out of context insult of a person which might force the person to leave a discussion or even a platform.
- *Dangerous speech*: This is a form of speech that attempts to fuel violence against a group by falsely portraying an imminent danger from them.
- *Fear speech*: Fear speech, as the name suggests, is a way to incite fear against a target community. This type of speech is usually long and argumentative in nature to make-believe the imminent existential fear of a community. It is highly effective and has the potential to push communities toward a physical conflict.
- *Cyberbullying*: Cyberbullying refers to repeated aggressive behaviour by an individual or a group toward a victim who cannot easily defend himself/herself.
- *Radicalization/extremism*: Online radicalization or extremism is a way to provoke terrorism and violence typically based on some nationalist agenda.

## 1.5 Medium of Hate Speech

In the online world hate speech is expressed in different forms. While the most studied form is text, other forms like images, audio and video are not uncommon. In fact, image and text are routinely combined to form *memes* which have become a regular means of sourcing hate speech into social media platforms. In the following we present a taxonomy of the different types of hate content and the contexts in which they are mostly studied.

- *Text based hate speech*: In its traditional form hate speech on online platforms are hurled using textual posts containing abusive and demeaning terms, slurs, swear words expressed against an individual or a community. Sources include posts on X, Facebook, Gab, WhatsApp etc.
- *Image based hate speech*: Images are often used to convey extremely hateful content over different social media platforms as tend to have a more intense impact on the victim(s). Sources include Instagram, Pinterest etc.
- *Audio based hate speech*: Recently, an emerging trend in spreading hate content is through podcasts many of which are controlled by alt-right and extremist groups. Even high profile Google podcasts have been shown to exhibit such hateful content.[13]
- *Video based hate speech*: One of the most intense forms of hateful content comes as videos, especially shorts and reels. While the typical hosting platforms for hate videos are BitChute, 4Chan etc., recent times have seen a surge in TikTok and Instagram reels and Youtube shorts spreading a wide variety of hate content.
- *Multi-modal hate speech*: All the above modals are combined at different proportions to create multi-modal hate content. Multi-modal content like memes and cartoons have become commonplace sources of hate speech and are abundantly seen across multiple platforms including X, Facebook, Google and Bing images, Instagram, WhatsApp, Telegram etc.

## 1.6 Contexts of Hate Speech

The genesis and the context of hate speech are strongly tied to the demography, language, culture and the platform under consideration [6]. However, there are other nuanced forms of contextualisation that we shall discuss in brief in this section.

### 1.6.1 Boundaries of Hate Speech and Intersection with Free Speech

The boundary between hate speech and free speech is very subtle. There are at least three different debatable views on this point. These are as follows.

- *Free speech absolutist*: The advocates of this approach contend that hate speech should not be suppressed unless it is hurled toward an individual that immediately results in a breach of peace.
- *Egalitarian approach*: The advocates of this approach contend that in order to do justice to the equality principle hate speech in all forms should be prohibited even in the absence of any explicit target.

---

[13] https://www.nytimes.com/2021/03/25/arts/google-podcasts-extremism.html.

- *Liberal approach*: The advocates of this approach lie somewhere in the middle of the above two (extreme) approaches. They contend that measures should be taken to proscribe speech that is specifically targeted to vilify a victim based on his/her race, gender, religion, ethnic origin, sexual orientation, or other sensitive attributes.

### 1.6.2 Focused Groups

Another type of contextualisation happens in the form of the focused groups that are repeatedly exposed to hate speech historically. Some of these forms (not exhaustive) are as follows.

- *Misogyny*: Misogyny is hatred expressed against women and girls. In the online world there have been multiple instances of such hatred including those in (i) the manosphere in Reddit [7], (ii) X during US presidential election [8], (iii) news media [9] (iv) rap and pop music [10], (v) video games [11], etc.
- *Sinophobia*: Sinophobia refers to anti-Chinese sentiment and hatred toward or fear of Chinese people. The online world underwent an infodemic with the outbreak of COVID-19. The largest number of victims of this infodemic were either people from mainland China or those settled in foreign countries [12].
- *Islamophopia*: Islamophobia refers to prejudice toward or fear of people who identify as Muslims [13]. Various fear-mongering agendas against Muslims have been perpetuated in mainstream social media as well as in fringe/extremist platforms. Some of the notable ones include the 'Great Replacement' conspiracy [14], COVID-19 outbreak conspiracy [15], 'Love-Jihad' conspiracy [16] etc.
- *Anti-semitism*: Anti-semitism refers to hostility and prejudice against the Jews. There is an abundance of research works evidencing anti-semitism in the online world [17, 18] along with conspiracy theories around homosexualisation [19], COVID-19 outbreak [20] etc.
- *Racism*: Racism refers to discrimination, hatred, or prejudice toward people or communities belonging to a specific racial or ethnic group. Racial differences have served as breeding grounds for hate speech for long. For instance, African Americans [21] and Asian Americans Zhang et al. [22] have been victims of racial hatred over generations. The impact of racial hatred has touched issues like politics [23], education [24], COVID-19 pandemic [25] and even online gaming [26].
- *Casteism*: Casteism refers to hatred or discrimination of a person or a group based on his/her caste. Violence and hatred related to caste have been plaguing the online and the physical world for a long. The discrimination and violence faced by Dalits [27] in India or the Osus [28] in Nigeria paint a vivid picture of caste-based proliferation of hatred.

In addition to the above, there are other forms of hatred based on physical disability, nationality, etc.

### 1.6.3 Emerging Communities

Multiple communities have emerged with a dedicated effort to perpetuate hate in social media. Two of the most notable ones are those related to genetic testing/genetic purity [29] and QAnon [30]. The former community attempts to use genealogy and ancestry reports that are easily available these days to establish their genetic 'superiority/purity' over other (minority) communities. The community is spread over multiple social media and video hosting platforms like Reddit, 4chan, etc. where there is a burgeoning growth in the discussion related to the importance of genetic testing and a parallel rise in the volume of toxic comments that are misogynistic, anti-semitic and racist in nature [29]. The latter community corresponds to what is better known as the 'super-conspiracy' theory, wherein a user with the nickname Q started creating multiple discussion threads on the /pol/ board of 4chan stating that numerous US government officials are involved in a satanic paedophile racket.[14] Soon the conspiracy gained momentum spreading like wildfire across other social media platforms and further claiming the Donald Trump was actively working against this racket to subdue them in all forms.

## 1.7 What to Expect from the Rest of This Book

The detailed mandate of the UN listed in one of the previous sections underscores the importance of the problems and the severity of the damage that could be potentially caused by hate events both in the online as well as the physical world. This book outlines step by step the different approaches that one could take to assist the UN in its plan of action using the modern tools of AI and machine learning. Broadly the book attempts to answer the following questions.

- *UN Key Commitment: Monitoring and analysing hate speech*: How does hate speech spread in the online world? Can one comment on the speed and depth of this spread using computational approaches? What are the long-lasting impacts of such spread?
- *UN Key Commitment: Addressing the root causes/drivers/technology*: What could be the first step to handle this issue? Can we detect hate speech using computer algorithms? Can the detection results obtained from the model be explained? Are there biases in the evaluation of these results? If there are, what are the nature and form of such biases?
- *UN Key Commitment: Countering hate speech*: How does one contain online hate? How does one resolve conflicts with the doctrine of freedom of speech? Can one use more speech to counter hate speech (aka counterspeech)? Is counterspeech generic or specific to target communities? If counterspeech is indeed effective, can one use technology to automatically generate counterspeech?

---

[14] https://www.thehindu.com/news/international/the-hindu-explains-how-qanon-went-from-a-bizarre-conspiracy-theory-to-an-election-talking-point/article32881282.ece.

## References

1. Zeerak Waseem, Thomas Davidson, Dana Warmsley, and Ingmar Weber. Understanding abuse: A typology of abusive language detection subtasks. In Zeerak Waseem, Wendy Hui Kyong Chung, Dirk Hovy, and Joel Tetreault, editors, *Proceedings of the First Workshop on Abusive Language Online*, pages 78–84, Vancouver, BC, Canada, August 2017. Association for Computational Linguistics.
2. Paul Rottger, Bertie Vidgen, Dirk Hovy, and Janet Pierrehumbert. Two contrasting data annotation paradigms for subjective NLP tasks. In Marine Carpuat, Marie-Catherine de Marneffe, and Ivan Vladimir Meza Ruiz, editors, *Proceedings of the 2022 Conference of the North American Chapter of the Association for Computational Linguistics: Human Language Technologies*, pages 175–190, Seattle, United States, 2022. Association for Computational Linguistics.
3. Paula Fortuna and Sérgio Nunes. A survey on automatic detection of hate speech in text. *ACM Computing Surveys (CSUR)*, 51(4):1–30, 2018b.
4. Jana Papcunová, Marcel Martončik, Denisa Fedáková, Michal Kentoš, Miroslava Bozogáňová, Ivan Srba, Robert Moro, Matúš Pikuliak, Marián Šimko, and Matúš Adamkovič. Hate speech operationalization: a preliminary examination of hate speech indicators and their structure. *Complex Intell. Syst.*, 2021.
5. Karsten Müller and Carlo Schwarz. *Fanning the Flames of Hate: Social Media and Hate Crime*. SSRN, 2020.
6. Michael Herz and Péter Molnár. *The content and context of hate speech: Rethinking regulation and responses*. Cambridge University Press, 2012.
7. Tracie Farrell, Miriam Fernandez, Jakub Novotny, and Harith Alani. Exploring misogyny across the manosphere in reddit. In *Proceedings of the 10th ACM conference on web science*, pages 87–96, 2019.
8. Nir Grinberg, Kenneth Joseph, Lisa Friedland, Briony Swire-Thompson, and David Lazer. Fake news on twitter during the 2016 us presidential election. *Science*, 363(6425):374–378, 2019.
9. Karla Mantilla. Gendertrolling: Misogyny adapts to new media. *Feminist studies*, 39(2):563–570, 2013.
10. Ronald Weitzer and Charis E Kubrin. Misogyny in rap music: A content analysis of prevalence and meanings. *Men and masculinities*, 12(1):3–29, 2009.
11. Keiko M McCullough, Y Joel Wong, and Natalie J Stevenson. Female video game players and the protective effect of feminist identity against internalized misogyny. *Sex Roles*, 82:266–276, 2020.
12. Leonard Schild, Chen Ling, Jeremy Blackburn, Gianluca Stringhini, Yang Zhang, and Savvas Zannettou. " go eat a bat, chang!": An early look on the emergence of sinophobic behavior on web communities in the face of covid-19. *arXiv preprint* arXiv:2004.04046, 2020.
13. Imran Awan. Islamophobia and twitter: A typology of online hate against muslims on social media. *Policy & Internet*, 6(2):133–150, 2014.
14. Milan Obaidi, Jonas Kunst, Simon Ozer, and Sasha Y Kimel. The "great replacement" conspiracy: How the perceived ousting of whites can evoke violent extremism and islamophobia. *Group Processes & Intergroup Relations*, 25(7):1675–1695, 2022.
15. Hussein Saadi Mohammad Ali and Yuli Rahmawati Mutiah. Islamophobia and conspiracy against muslim during covid-19 outbreak in india. *Journal of Interdisciplinary Islamic Studies*, 1(1):9–16, 2022.
16. Eviane Leidig. From love jihad to grooming gangs: Tracing flows of the hypersexual muslim male through far-right female influencers. *Religions*, 12(12):1083, 2021.
17. Monika Schwarz-Friesel and Jehuda Reinharz. *Inside the antisemitic mind: the language of Jew-Hatred in contemporary Germany*. Brandeis University Press, 2017.

# References

18. Mohit Chandra, Dheeraj Pailla, Himanshu Bhatia, Aadilmehdi Sanchawala, Manish Gupta, Manish Shrivastava, and Ponnurangam Kumaraguru. "subverting the jewtocracy": Online anti-semitism detection using multimodal deep learning. In *Proceedings of the 13th ACM Web Science Conference 2021*, pages 148–157, 2021.
19. Kristoff Kerl. The 'conspiracy of homosexualisation': Homosexuality and anti-semitism in the united states, 1970s–1990s. *Journal of Modern European History*, 20(3):352–370, 2022.
20. Rakib Ehsan et al. Weaponising covid-19: Far-right antisemitism in the united kingdom and united states. *Henry Jackson Society, May*, 2020.
21. Thomas Davidson, Debasmita Bhattacharya, and Ingmar Weber. Racial bias in hate speech and abusive language detection datasets. In *Proceedings of the Third Workshop on Abusive Language Online*, pages 25–35, 2019.
22. Yan Zhang, Lening Zhang, and Francis Benton. Hate crimes against asian americans. *American Journal of Criminal Justice*, pages 1–21, 2021.
23. Edward L Glaeser. The political economy of hatred, 2002.
24. Michalinos Zembylas. The affective politics of hatred: Implications for education. *Intercultural Education*, 18(3):177–192, 2007.
25. Stephen M Croucher, Thao Nguyen, and Diyako Rahmani. Prejudice toward asian americans in the covid-19 pandemic: The effects of social media use in the united states. *Frontiers in Communication*, 5:39, 2020.
26. Stephanie M Ortiz. "you can say i got desensitized to it": How men of color cope with everyday racism in online gaming. *Sociological Perspectives*, 62(4):572–588, 2019.
27. Nidhi Sadana Sabharwal and Wandana Sonalkar. Dalit women in india: At the crossroads of gender, class, and caste. *Global Justice: Theory Practice Rhetoric*, 8(1), 2015.
28. Nneka Sophie Amalu, Yusuf Abdullahi, and Ekong Demson. Caste conflict in nigeria: The osu/diala experience in igboland, 1900-2017. *Global Journal of Social Sciences*, 20(1):77–85, 2021.
29. Alexandros Mittos, Savvas Zannettou, Jeremy Blackburn, and Emiliano De Cristofaro. "and we will fight for our race!" a measurement study of genetic testing conversations on reddit and 4chan. In *Proceedings of the International AAAI Conference on Web and Social Media*, volume 14, pages 452–463, 2020.
30. Antonis Papasavva, Jeremy Blackburn, Gianluca Stringhini, Savvas Zannettou, and Emiliano De Cristofaro. "is it a qoincidence?": An exploratory study of qanon on voat. In *Proceedings of the Web Conference 2021*, pages 460–471, 2021.

# Analysis   2

In this chapter, we shall discuss the different works that attempted to analyse hate content (and its variants) across different social media platforms. We present works that study both moderated platforms like X and loosely moderated platforms like Gab. We further discuss some of the effects of hate speech in the online as well as the physical world.

## 2.1 Spread of Hate Speech Across Online Platforms

In the following, we shall describe some of the popular works that analyse the spread of hate content across different platforms. These platforms can be subdivided into two broad categories—platforms with moderation and platforms with lax/loose or no moderation. Examples of the former include X, Facebook and YouTube which have strict hate speech policies in place and stringent moderation pipelines that run on a workforce of both human workers and sophisticated algorithms.

### 2.1.1 Moderated Platforms

In the following, we discuss some of the studies related to hateful content and users on platforms with strict hate speech moderation policies in place.

**Hateful Users on X**

One of the early interesting studies on moderated platforms was done by Ribeiro et al. [1] where the authors studied the activities of hateful users on X. They proposed a methodology to gather and annotate hateful users. The annotation does not merely use a lexicon-based approach but relies on the inspection of the entire profile of the users. The authors used the

CrowdFlower platform to annotate 4972 users as hateful or not. The annotation resulted in 544 hateful users and 4,428 normal ones. The authors made quite a few interesting observations based on the analysis of this data. First, they found that the creation date of hateful users is on average later than the normal ones. This can be attributed to the fact that hateful users quickly get suspended due to the strict moderation policy of X and have to join the network again and again each time with new account credentials. The authors further found that the hateful users behave like 'power users'—they post more tweets and within short intervals, favourite other people's tweets more and follow more users on the platform. Moreover, they do not behave like spammers. The hateful users were observed to be more centrally situated in the network and were densely connected among themselves. In fact, the density of retweets among hateful users was found to be alarmingly high—a hateful user was observed to be 71 times more likely to retweet another hateful user.

**Islamophobic Attacks in Western Countries**

In Olteanu et al. [2], the authors collected data from three sources—Twitter, Reddit and news articles[1] to study online hate speech following (a) Islamist terrorist attacks and (b) Islamophobic attacks perpetrated in Western countries. The dataset spans 19 months—from January 1, 2016, to August 1, 2017—and covers messages potentially related to hate and counterspeech. The authors made certain highly interesting observations from this analysis. After an Islamist terrorist attack, terms related to violence and offence increased while there was no such increase observed after a hate crime committed against an individual or a group of Muslims. On the other hand, terms with counterspeech increased after terrorism but not after hate crime. Thus in both situations, hate crimes committed toward Muslims did not receive social media attention while terrorist attacks by Muslims were always in the limelight.

**Psychological Effects of Online Hate Speech**

The authors in Saha et al. [3] studied online hate speech in a dataset of 6 million Reddit comments shared in 174 college communities. The prevalence of hate speech was quantified by a novel index proposed by the authors of College Hate Index (CHX). To define this metric the authors used a manually curated hate lexicon of 157 keywords (words/phrases). The CHX of an online college community is defined as the ratio of the normalized occurrence of the hate keywords in the college subreddit to the same measure in banned subreddits (which were banned due to the presence of extreme hateful content in them and act as an upper bound for this measure). Thus one has

$$CHX_T(S) = P_T(S)/P_T(B) \tag{2.1}$$

where $S$ is the college subreddit, $B$ is the extreme hateful banned subreddit, $P_T(S)$ and $P_T(B)$ are the normalized occurrence of the hate keywords in $S$ and $B$. $T$ refers to whether

---

[1] http://gdeltproject.org/.

CHX is measured for a specific hate category (e.g., race, gender, ethnicity, religion etc.) or all categories taken together.

The study of CHX for college groups resulted in some very alarming observations. The authors found there is an abundance of hate speech in college subreddits and around 25% of them show higher hate speech content compared to non-college subreddits. The authors further noted that exposure to hate led to increased stress levels among victims. Nevertheless, not all victims were equally affected by the exposure to hate speech. Individuals with lower levels of psychological endurance were found to be more vulnerable to hate content and subsequent emotional outbursts. This work laid the foundation for studying the psychological impacts of hateful speech in online communities in general and college campuses in specific.

### 2.1.2 Platforms with Lax Moderation

Many of the mainstream platforms like X, Facebook etc. have installed heavy moderation in view of the increasing threat manifesting from online hate speech. Therefore, alternative platforms like Gab,[2] 4Chan,[3] BitChute[4] etc. have emerged as new resorts for the suspended hateful users. These alternative platforms portray themselves as 'champions of free speech' and consequently have lax moderation. Under this camouflage such websites have reportedly welcomed alt-right and extremist users who were oftentimes banned by the mainstream social media platforms.[5] In the following we shall discuss some of the works that analysed hateful content on such platforms.

**Gab**

The authors in Zannettou et al. [4] crawled https://gab.com between August 2016 and January 2018 to obtain 22M posts from 336k users. Some of the most frequent hashtags on the platform was found to be #Alt-right, #BanIslam. The authors in Mathew et al. [5] built on this work and studied the cascade properties of the content spread by hateful and non-hateful users. They started by manually labelling a small set of hateful users and then used a diffusion model to estimate the extent of the hatefulness of the neighbours of these users. First, they built a repost network where each node is a user and a directed edge from user $u_i$ to user $u_j$ with weight $w_{ij}$ indicates that $u_i$ has reposted $w_{ij}$ messages of $u_j$ (see Fig. 2.1a). This is converted to a belief network as shown in Fig. 2.1b. This is done as follows—user B in the figure reposts 5 messages from user A and herself posts 15 messages; the belief of user B therefore on user A is $\frac{5}{(5+15)} = 0.25$ and on herself is $\frac{15}{(5+15)} = 0.75$. Naturally, the direction of the edges has been reversed in this network indicating the belief of a user on another. Next, the authors run the DeGroot model [6] on the belief network starting

---

[2] https://gab.com/.
[3] https://www.4chan.org/.
[4] https://www.bitchute.com/.
[5] https://www.adl.org/blog/when-twitter-bans-extremists-gab-puts-out-the-welcome-mat.

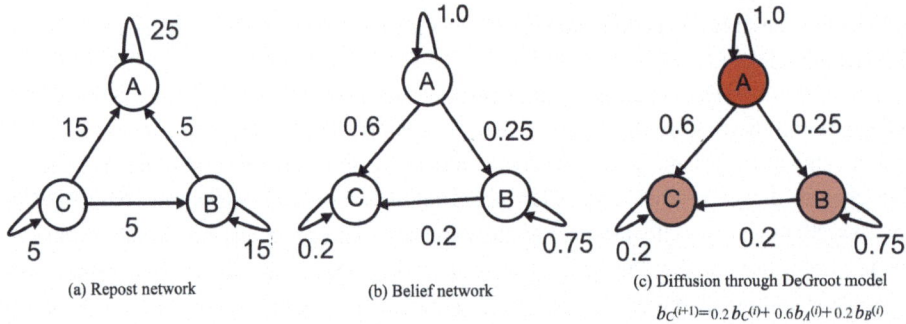

**Fig. 2.1** Identification of hateful users based on the DeGroot model of diffusion. Node A marked in dark red indicates the manually labelled hateful user. The nodes B and C marked in light red indicate the extent of hatefulness inferred by the diffusion model at convergence

from manually labelled hateful nodes (see Fig. 2.1c). These starting nodes were assigned a hatefulness score of 1 while all other nodes were assigned a hatefulness score of 0. Next, the hatefulness scores were updated iteratively till the point where there was no further change in them. As is shown in Fig. 2.1c, the hatefulness scores of node A (manually labelled as hateful), B and C in the start step are $b_A^{(0)} = 1.0$, $b_B^{(0)} = 0.0$ and $b_C^{(0)} = 0.0$ respectively. In the next step, the score of node C would be updated to $b_C^{(1)} = 0.2 b_C^{(0)} + 0.6 b_A^{(0)} + 0.2 b_B^{(0)} = 0.2 \times 0.0 + 0.6 \times 1.0 + 0.2 \times 0.0 = 0.6$. Upon convergence, all the nodes in the network would have a hatefulness score between 0 and 1. Nodes having scores in the range [0.75, 1] were identified as hateful users (KH (*known-hateful*), 2,290 in number) while those having scores in the range [0, 0.25] were identified as non-hateful (NH (*non-hateful*), 58,803 in number) users. The authors demonstrated through human judgement experiments that the algorithm was highly accurate in identifying hateful users.

Once the nodes got labelled (KH or NH) the authors studied information cascades based on them. To this purpose, they constructed a variant of the repost network where nodes were again users while edges indicated followership relations (see Fig. 2.2a). Each node is has a number on it representing the time of repost of a message by the node. The hypothesis is that the time ordering would indicate which node unambiguously influences a particular node. However this is a non-trivial problem since the unambiguous influencer might be difficult to obtain. For example in Fig. 2.2a, it is unclear if the repost by C at time 150 is influenced by A or B (both of whom are in the receipt of the message before C and both of whom C follows). The authors used a heuristic approach called LRIF (least recent influencer) [7] to break this tie which assumes that the node which received the message least recently is the influencer (A in this case).[6] This heuristic reduces the network into a DAG (Fig. 2.2a) which can then be used to study the cascade properties.

---

[6] MRIF (most recent influencer) heuristic is the other alternative wherein B would be designated as the influencer of C.

## 2.1 Spread of Hate Speech Across Online Platforms

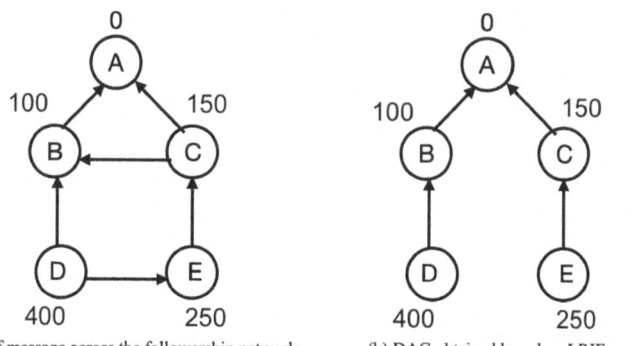

**Fig. 2.2** Cascade of information over the followership network

**Table 2.1** The different cascade properties for the posts initiated by KH and NH users

|         | Posts |      | Attachments |      | Topics |      |
|---------|-------|------|-------------|------|--------|------|
| Feature | KH    | NH   | KH          | NH   | KT     | NT   |
| Size    | **1.28** | 1.21 | **1.34** | 1.23 | **1.68** | 1.51 |
| Depth   | **0.13** | 0.09 | **0.16** | 0.11 | **0.30** | 0.24 |
| Breadth | **1.13** | 1.10 | **1.15** | 1.11 | **1.30** | 1.24 |
| AD      | **0.11** | 0.08 | **0.14** | 0.10 | **0.26** | 0.22 |
| SV      | **0.13** | 0.09 | **0.16** | 0.11 | **0.31** | 0.25 |

The authors constructed DAGs for each individual message initiated by a KH user and compared their aggregate properties with those for the NH users. Next, the authors studied the following standard metrics [8]—(i) *size*, i.e., number of unique users, (ii) *depth*, i.e., length of the largest path in the cascade, (iii) *average depth*, i.e., the average depth of each path from the root node, (iv) *breadth*, i.e., the maximum width of the cascade, and (v) *structural virality* which indicates viral nature of the post. The mean behaviour of the hateful versus non-hateful behaviour is shown in Table 2.1. The results are shown for (i) all posts (ii) multimedia-based posts only and (iii) posts in hate (KT) versus non-hate (NT) topics.[7] The authors showed that in all these categories the messages from KH users could reach out to a larger audience, were more viral and penetrated deeper and wider into the network compared to messages posted by NH users. The differences in the distribution of all these properties were found to be statistically significant.

The same authors studied the temporal evolution of hateful users and content on Gab in a follow-up work [9]. In this work, they showed that hateful users were more successful in quickly moving to prominent positions in the core of the network. Further, they found that

---

[7] KT: a higher proportion of hateful content; NT: those having a lower proportion of hateful content.

not only the amount of hate content on the platform has been increasing over time but also the overall linguistic style of even the normal users was getting aligned to the hateful users.

**4Chan**

In Papasavva et al. [10], the authors analyzed the /pol/ thread of the 4Chan platform based on the dataset crawled between June 29, 2016, to November 1, 2019. In order to identify hate content the authors used the Perspective API. They showed that around 37% of the content had some level of toxicity while 27% was found to be severely toxic. Various activist movements and alt-right political ideologies were also found to feature on the platform.

### 2.1.3 Other Forms of Harmful Content

Recently, most online platforms have installed heavy moderation policies and therefore perpetrators of hate messages are routinely suspended or banned. In order to bypass this newer forms of harmful content have emerged which do not contain toxic or abusive words thus making them harder to detect. In the following, we discuss some such forms of harmful content.

**Fear Speech**

In a recent work [11], the authors have identified a more nuanced form of harmful content—*fear speech*—that is posted by perpetrators to incite fear against a minority community. These posts are largely non-toxic in nature and are strung in an argument chain (often fake) which makes them more believable particularly to benign users. The authors showed that such posts are plentiful in public WhatsApp groups discussing political issues. Later [12] the authors also demonstrated the prevalence of such speech in Gab. The authors found that such posts enjoy more reshares and a longer lifetime. A large number of users spread them and a large number of groups get affected. More interestingly, certain emojis are used in groups along with a post to make the 'fear-effect' far-reaching. Further, the authors found that the average level of engagement of users with fear speech posts is much higher than hate speech posts. What is more worrisome is that normal users get mentioned more, reply more and repost more to fear speech compared to hate speech. The authors showed that even other platforms like X and Facebook contain fear speech in large proportions and some of these have remained unnoticed/undetected by the moderators since 2016.

**Dangerous Speech**

As defined by the authors in Alshehri et al. [13] dangerous speech refers to posts that attempt to inflict physical pain, injury, or damage on someone in retribution for something which may or may not have been done. In this work, the authors collected data from the Arabic X platform and annotated data based on the above definition. The authors found various interesting patterns of threats of physical harm which are enumerated below.

## 2.1 Spread of Hate Speech Across Online Platforms

- *Conditional threats*: The threat statement here is a conditional where the consequent corresponds to a physical threat to the victim and an antecedent corresponds to the deterrence of the victim (or a third person) from doing something. Example: 'I slaughter you if you (F) do anything.
- *Threats along with commands*: In this case, a command is present along with the threat. Example: 'I say get out before I hit your face'.
- *Threats along with questions*: Here the threat is expressed as a question statement. Example: 'Does it work if I rape you?'
- *Imperative threats*: These threats are expressed in deontic or epistemic modalities. Example 'May I rape you?!'
- *Threats as metaphors*: In this case, the threats are metaphorically expressed. Example: "I would like to tell my Manchester (football club) fans that we will rape them tomorrow'.
- *Emojis*: A lot of threats could be expressed by the use of emojis in the posts. Examples include: 'I torture you' (😱), 'I rape you' (👿), 'I hit you' (😡) etc.

Another recent work Dash et al. [14] attempted to identify dangerous speech users on X in the Indian sub-continent and characterize them. The authors first used a lexicon-based approach to identify dangerous tweets and subsequently refined this sample by manual annotation. The tweets corresponded to three events that were reported to account for a lot of violence in the physical world. The user set was built from the list of politician accounts plus the non-politician, non-NRI accounts (influencers) these politicians followed. The authors flagged a user from this list as a dangerous speech user if they made at least one post from the dangerous tweet sample curated earlier. Next, they built a network based on the retweets among these nodes. They then identified additional dangerous users by running a diffusion model on the retweet network of these users similar to Mathew et al. [5]. Subsequently, the authors showed that dangerous users were more active compared to the other users in the network. These users were also found to be more influential in terms of the volume of verified accounts and the total number of followings. The contents posted by dangerous users were viewed by more polarized users who are highly susceptible to incitement. Most of the dangerous speech users were identified as people from mass media or low-ranked politicians who acted as broadcasting agents promoting the steady growth and penetration of dangerous speech into the social network.

**Dog Whistles**

Dog whistles are a form of coded expressions that usually convey dual meaning. While one of these meanings is meant for a broad audience the other one, mostly hateful and inciting, is meant for a narrow in-group. Such codes are commonplace to bypass platform moderation as well as political/legal repercussions. For instance, while "international banks" might sound quite literal to most people, anti-Semites interpret it in a very different way. For them, this often refers to the supposed existence of a cabal of international Jewish bankers constantly working to undermine the US democracy.[8] In Mendelsohn et al. [15] the authors presented a large-scale study investigating the various properties of dog whistles. First, they identified a typology of dog whistles and curated the largest ever glossary of 300+ dog whistles. Along with each dog whistle, the glossary also has context information in which the dog whistle was used historically by one or more US politicians in their speeches. The authors categorized a dog whistle based on three factors—register, type and persona. Register refers to whether a dog whistle is formal/offline or informal/online. Formal/offline correspond to the dog whistles used in statements by political elites while informal/online correspond to those originating from the Internet/social media. Types can be either only persona signalling (Type I) or persona signalling with an added message (Type II). Finally, persona corresponds to the in-group for which the dog whistle is meant. The authors found that since the 1990s there has been a steady use of racial dog whistles and after 9/11 these were not only anti-Black but also Islamophobic and anti-Latinx.

Subsequently, the authors studied the power of LLMs (GPT-3) in identifying dog whistles. The LLM finds more difficulty in identifying informal/online registers compared to the formal/offline ones. Quite alarmingly, the LLM recall is $\leq 20\%$ for the informal/online register. On the other hand, the performance for the formal/offline register was quite high with a maximum of $\sim 100\%$ for the Islamophobic and a minimum of $\sim 45\%$ for the transphobic persona categories. Overall the authors demonstrated that harmful content camouflaged within dog whistles was efficiently able to bypass SOTA toxicity detection models thus calling for more research efforts in this area which is still very hazy.

### 2.1.4 Newly Emerging Platforms

While the spread of hate speech has inflicted mainstream social media platforms like X and Facebook for long, the advent of new platforms with creative content in the form of short videos has fuelled this spread to newer heights. One such platform with extremely lax moderation and heavily dominated by the young (in fact very young) population is TikTok. It is a video-sharing platform run by the Chinese company ByteDance and was launched worldwide in 2018. The uniqueness of this platform is that it can host only *short* videos of duration ranging from 3 s to 10 min. From its inception, this platform (and many of its clones) has reportedly been a breeding ground for propagating racist, sexist, and homophobic language. The weak moderation and privacy policies make it easy for the perpetrators to spread violence through hate videos and inflammatory narratives. In Abdullah [16] the author studied TikTok to show how it swayed the Malaysian election via hate speech propaganda. In particular, the author analyzed 679 popular TikTok videos with over 1,000 views and found 373 of them containing hateful narratives and propaganda. For instance, on November 11, 2022, a TikTok user (@125cc_madi) shared a video that showed DAP supporters criticising Muslim PAS "stupid Muslim ulama." This single video went viral receiving as high as 16k

---

[8] https://www.vox.com/the-big-idea/2016/11/7/13549154/dog-whistles-campaign-racism.

views. Similarly, videos with hashtag #13mei on TikTok infuriated the Malaysians during the latest election, accusing political actors of exploiting the 1969 incident of a conflict between Malay and Chinese communities in Kuala Lumpur to arouse anti-China sentiment.

Another study [17] showed the rise of *incels* on the TikTok platform. Incels are involuntary celibates and a subgroup of the manosphere who have become notorious due to their associations with an array of violent attacks. The authors showed that on TikTok the incel groups work undercover. Emotional appeals and pseudo-science are used in abundance to spread the incelospehre ideology. The authors further showed that these interconnected groups of incels have been able to promote the normalisation of blackpilling beliefs, reinforce misogyny and consequently, justify the rape culture.

Recently, the think-and-do-tank ISD set forth a mission to study [18] hate and extremism on the TikTok platform. They collected a sample of 1030 videos spanning eight hours of content to perform this analysis. They found 312 videos promoting white supremacy and upholding the Great Replacement and white genocide conspiracy theories. They further identified 246 videos that praised and glorified extremist figures/organisations like Adolf Hitler, ISIS, Ratko Mladic or Oswald Mosley. Around 26 video posts denied the existence of the Holocaust. Further around 279 videos used music and video creation features to spread extremist content on TikTok. The most viewed video in the data sample had 2 million views meant to spread anti-Asian hatred linked to COVID-19. Among the top 10 most viewed videos were those that denied the Bosnian genocide and the Holocaust with 655,800 and 233,000 views respectively. The hateful content creators were found to successfully use the algorithmic promotion tools of the platform to enhance the spread of extremist and hate content. They were also able to evade the (weak) moderation policies of the platform by either whitewashing or strategically using the privacy functions and comment restrictions or by slightly changing the hashtag spellings. Finally, the study found that hateful extremist contents were removed from the platform inconsistently and as high as 81.5% of the original hateful sample remained on the platform when it was rechecked at a later date.

## References

1. Manoel Horta Ribeiro, Pedro H Calais, Yuri A Santos, Virgílio AF Almeida, and Wagner Meira Jr. Characterizing and detecting hateful users on twitter. In *Twelfth International AAAI Conference on Web and Social Media*, 2018.
2. Alexandra Olteanu, Carlos Castillo, Jeremy Boy, and Kush Varshney. The effect of extremist violence on hateful speech online. In *Proceedings of the international AAAI conference on web and social media*, volume 12, 2018.
3. Koustuv Saha, Eshwar Chandrasekharan, and Munmun De Choudhury. Prevalence and psychological effects of hateful speech in online college communities. In *Proceedings of the 10th ACM conference on web science*, pages 255–264, 2019.
4. Savvas Zannettou, Barry Bradlyn, Emiliano De Cristofaro, Haewoon Kwak, Michael Sirivianos, Gianluca Stringini, and Jeremy Blackburn. What is gab: A bastion of free speech or an alt-right

echo chamber. In *Companion of the The Web Conference 2018 on The Web Conference 2018*, pages 1007–1014, 2018.

5. Binny Mathew, Ritam Dutt, Pawan Goyal, and Animesh Mukherjee. Spread of hate speech in online social media. In *Proceedings of the 10th ACM conference on web science*, pages 173–182, 2019a.

6. Benjamin Golub and Matthew O Jackson. Naive learning in social networks and the wisdom of crowds. *American Economic Journal: Microeconomics*, 2 (1): 112–49, 2010.

7. Eytan Bakshy, Jake M Hofman, Winter A Mason, and Duncan J Watts. Everyone's an influencer: quantifying influence on twitter. In *Proceedings of the fourth ACM international conference on Web search and data mining*, pages 65–74. ACM, 2011.

8. Soroush Vosoughi, Deb Roy, and Sinan Aral. The spread of true and false news online. *Science*, 359 (6380): 1146–1151, 2018. ISSN 0036-8075.

9. Binny Mathew, Anurag Illendula, Punyajoy Saha, Soumya Sarkar, Pawan Goyal, and Animesh Mukherjee. Hate begets hate: A temporal study of hate speech. *Proceedings of the ACM on Human-Computer Interaction*, 4 (CSCW2): 1–24, 2020a.

10. Antonis Papasavva, Savvas Zannettou, Emiliano De Cristofaro, Gianluca Stringhini, and Jeremy Blackburn. Raiders of the lost kek: 3.5 years of augmented 4chan posts from the politically incorrect board. In *Proceedings of the international AAAI conference on web and social media*, volume 14, pages 885–894, 2020.

11. Punyajoy Saha, Binny Mathew, Kiran Garimella, and Animesh Mukherjee. " short is the road that leads from fear to hate": Fear speech in indian whatsapp groups. *arXiv preprint* arXiv:2102.03870, 2021.

12. Punyajoy Saha, Kiran Garimella, Narla Komal Kalyan, Saurabh Kumar Pandey, Pauras Mangesh Meher, Binny Mathew, and Animesh Mukherjee. On the rise of fear speech in online social media. *Proceedings of the National Academy of Sciences*, 120 (11): e2212270120, 2023.

13. Ali Alshehri, Muhammad Abdul-Mageed, et al. Understanding and detecting dangerous speech in social media. In *Proceedings of the 4th Workshop on Open-Source Arabic Corpora and Processing Tools, with a Shared Task on Offensive Language Detection*, pages 40–47, 2020.

14. Saloni Dash, Rynaa Grover, Gazal Shekhawat, Sukhnidh Kaur, Dibyendu Mishra, and Joyojeet Pal. Insights into incitement: A computational perspective on dangerous speech on twitter in india. In *ACM SIGCAS/SIGCHI Conference on Computing and Sustainable Societies (COMPASS)*, pages 103–121, 2022.

15. Julia Mendelsohn, Ronan Le Bras, Yejin Choi, and Maarten Sap. From dogwhistles to bullhorns: Unveiling coded rhetoric with language models. In Anna Rogers, Jordan Boyd-Graber, and Naoaki Okazaki, editors, *Proceedings of the 61st Annual Meeting of the Association for Computational Linguistics (Volume 1: Long Papers)*, pages 15162–15180, Toronto, Canada, July 2023. Association for Computational Linguistics. URL https://aclanthology.org/2023.acl-long.845.

16. Nurul Azmira Binti Abdullah. *SOCIAL MEDIA AND ELECTION CAMPAIGN: TIKTOK AS 2022 MALAYSIAN GENERAL ELECTION BATTLEGROUND*. PhD thesis, International Islamic University Malaysia, 2023.

17. Anda Iulia Solea and Lisa Sugiura. Mainstreaming the blackpill: Understanding the incel community on tiktok. *European Journal on Criminal Policy and Research*, 29 (3): 311–336, 2023.

18. Ciaran O'Connor. Hatescape: An in-depth analysis of extremism and hate speech on tiktok. *Institute for Strategic Dialogue*, 24, 2021.

# Detection

This chapter will discuss the techniques developed over time for hate speech detection. We will explore how hate speech detection techniques have evolved, ranging from keyword-based methods and machine learning techniques to deep learning models and, currently, utilizing generative AI-based large language models such as ChatGPT. We formulate the hate speech detection task as follows:

Given a dataset **D** consisting of pairs $(X, Y)$, where $X = \{w_1, w_2, \ldots, w_m\}$ represents a text sample consisting of a sequence of words, and **Y** represents its corresponding label, the goal is to learn a classifier $F : F(X) \rightarrow Y$ that can accurately predict the presence or absence of hate speech in unseen text samples, where $Y \in \{y_1, y_2, \ldots, y_n\}$ is the ground-truth label. Here, $m$ represents the number of words present in a text, and $n$ represents the number of classification labels. If the task is only to decide if a given post is hate speech or not, then it becomes a binary classification. If we have more than two classes, it becomes a multi-class classification. Figure 3.1 shows the pipeline of the general hate speech classification task.

**Dataset preparation**: This step involves collecting and cleaning a dataset consisting of examples of hate and non-hate speech. The dataset should be representative of the type of hate speech that the model will be deployed to detect. In addition, the dataset should be cleaned to remove any noise or irrelevant information. Splitting the dataset into training, validation, and test sets is essential. The training set will be used to train the model, the validation set will be employed to fine-tune the model parameters, and the test set will be utilized to evaluate the model's performance on unseen data.

**Feature extraction and development**: Once the dataset is prepared, the next step is to extract features from the data. These features can be lexical, syntactic, semantic, or social media-specific [2]. Lexical features include things like the words used in the text, the length of the text, and the number of exclamation points and question marks. Syntactic

**Fig. 3.1** A comprehensive framework for identifying hate speech [1]. The different parts are indicated in different colors

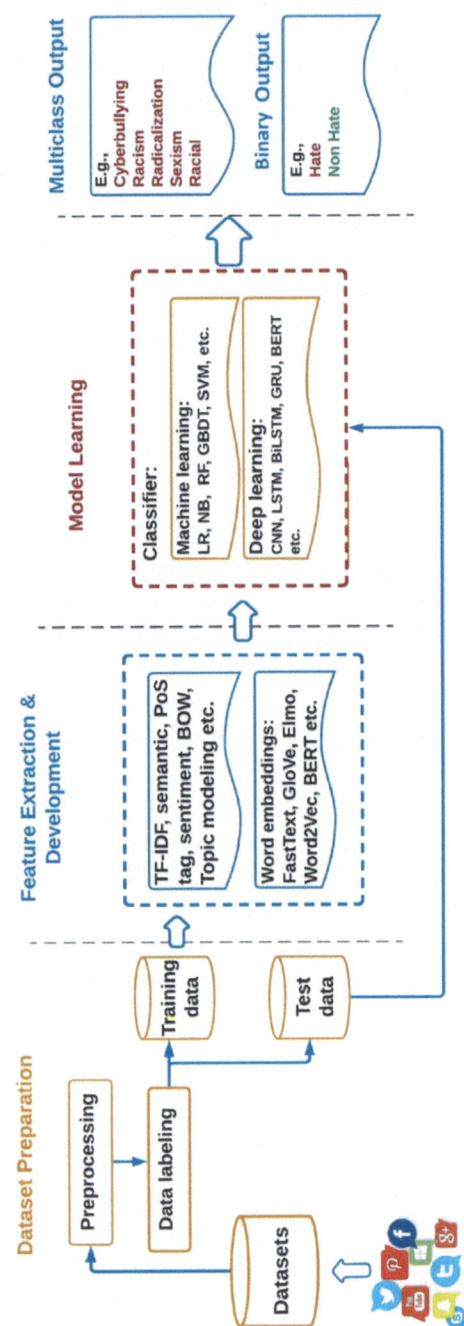

features include things like the part-of-speech tags of the words and the sentence structure. Semantic features include things like the text's sentiment and the presence of certain keywords. Social media-specific features include the number of likes and retweets on a post. The type of features that are extracted will depend on the specific task and the availability of resources. For example, suppose the goal is to build a simple hate speech detection model. In such a case, it may be sufficient to use only lexical features. However, the goal maybe to build a more complex model to detect more nuanced hate speech types. In that case, it would be necessary to use a combination of lexical, syntactic, semantic, and social media-specific features.

**Model learning**: Once the features have been extracted, the next step is to train a hate speech detection model on the extracted features. The type of model used will depend on the specific task and the availability of resources. Some standard machine learning models that can be used for hate speech detection include logistic regression [3], naïve Bayes [4], random forests [5], etc. These models are relatively simple to train and can effectively detect hate speech. Deep learning [6] models can also be used for hate speech detection. While deep learning models are more complex to train, they can be more effective at detecting hate speech and incredibly more nuanced types of hate speech.

**Multi-class output**: The hate speech detection model can be trained to produce either a binary label (e.g., "hate speech" or "not hate speech") or multiple categories (e.g., "cyberbullying", "racism", "radicalization", etc.). Once trained, the model is capable of analyzing new text and determining whether it qualifies as hate speech.

While we get a comprehensive understanding of the hate speech detection pipeline, it is crucial to note that not all detection systems follow the same architecture. Now, we shall navigate through various models existing in the literature shedding light on their strengths, limitations, and real-world applications.

## 3.1 Keyword-Based Techniques

Keyword-based hate speech detection is an early and straightforward method to identify hate speech in online content. It involves creating a lexicon of offensive keywords associated with hate speech. If any of these keywords appear in a post, it is considered to potentially contain hate speech [7]. The lexicon is crafted by experts, language specialists, or through automated methods and includes offensive words related to various characteristics. Once established, the lexicon is used to scan and analyze content for these keywords, either manually or through automated text analysis. If any hateful keywords are found, the content is flagged as potential hate speech, and further actions can be taken, such as content moderation or reporting.

While the keyword-based approach seems straightforward, it has several limitations as follows.

- First, the accuracy of hate speech detection relies solely on the presence of specific keywords. This means that if a post does not include any of the lexicon's keywords, it may not be detected as hate speech, even if it contains harmful or discriminatory content, resulting in false negatives [8].
- Conversely, the presence of hateful keywords does not always indicate hate speech. Context is crucial in understanding the true intention behind a post. For instance, a post mentioning a racial slur in a discussion condemning its usage might not be promoting hate speech but instead discussing it critically. This introduces the issue of false positives, where content is mistakenly flagged as hate speech due to the presence of specific keywords, even though the actual intent may not be hateful [9].
- Another challenge is that language constantly evolves, and new forms of hate speech emerge over time [10]. This requires the continuous updating and maintenance of the lexicon to ensure its effectiveness. In addition, hate speech can often be expressed using subtle or coded language that a keyword-based approach may not capture [11].

Despite these limitations, keyword-based hate speech detection can still be useful when combined with other techniques and approaches. It can serve as an initial filter to identify potentially problematic content that warrants further examination. However, relying solely on this approach cannot accurately detect and address the complex and dynamic nature of hate speech on online platforms.

## 3.2 Machine Learning Techniques

Machine learning-based hate speech detection has recently gained significant attention as a powerful approach to identify and combat hate speech in online platforms automatically [2]. The process of developing a machine learning-based hate speech detection system involves a set of key steps. Initially, a labeled dataset containing hate and non-hate speech samples is needed. The dataset is crucial for training the models, allowing them to learn hate speech patterns. After preparing and splitting the dataset into train and test sets, the machine learning models are trained using features extracted from the text data.

### 3.2.1 Extracting Meaningful Features

These features are characteristics of the text that help the model distinguish between hate and non-hate speech. Common features [2, 12, 13] include one or more of the following.
**Lexical features**:

- *Word frequencies* represent the count or frequency of specific words, capturing the presence of hateful or discriminatory language [14].

## 3.2 Machine Learning Techniques

- *N-grams* consider sequences of words, capturing contextual information and potentially revealing patterns associated with hate speech [13].
- *Syntactic features* focus on the grammatical structure of the text, identifying patterns such as the use of personal pronouns, verbs, or adjectives that may indicate hateful intent [2].
- *Semantic representations* aims to capture meaning and context, leveraging techniques like word embeddings or topic modelling.

**Word embeddings**:

- *Word2Vec* [12] embedding represents individual words as vectors, capturing their semantic relationships and contextual usage.
- *Doc2Vec* [15] embedding assigns unique vector representations to entire documents, allowing the model to capture the context and semantic similarity between different texts.
- *LASER* [16] embedding combines the context of neighbouring words to generate embeddings, considering the order of words in a sentence.

Incorporating these advanced embedding information directly as features enables machine learning models to gain a deeper understanding of the text data, identifying subtle nuances and hidden patterns associated with hate speech [17]. This approach helps overcome the limitations of hand-crafted features and improves accuracy and performance in hate speech detection.

### 3.2.2 Popular Algorithms

Traditional machine learning models, including logistic regression [3], SVMs [18], naïve Bayes [4], decision trees [19], random forest [5], and K-nearest neighbors [20], have been widely used for hate speech detection. Here is a brief explanation of how some popular models are employed for this task.

- **Logistic regression** (LR) [3] models estimate the probability of an instance belonging to a specific class based on a linear combination of the features.
- **Support vector machines** (SVM) [18] aims to find an optimal hyperplane that separates different classes in a high-dimensional feature space.
- **Decision trees** [19] are employed to create a hierarchical structure of decisions based on input features, aiding in the identification of patterns associated with hate speech.
- **K-nearest neighbors** (KNN) [20] makes predictions based on the majority class of its $k$-nearest neighbors.

The trained models are evaluated using the test set, which contains instances the models have not seen during training. Performance metrics are calculated to assess the effectiveness of the models in correctly classifying hate from non-hate speech instances. Furthermore, strategies like cross-validation and hyperparameter tuning may be employed to ensure robustness and optimize the model's performance.

### 3.2.3 Limitations

Traditional machine learning models have their limitations. They heavily rely on handcrafted features, which may not capture the nuances of complex text data. They might struggle with generalization to unseen or evolving types of hate speech. Moreover, the dynamic nature of online platforms demands continuous updates to adapt to these changes, making traditional models less scalable and efficient [8]. Nonetheless, these approaches have demonstrated promising results in hate speech detection and have served as a foundation for further advancements in the field. They provide a solid baseline for comparison with more advanced techniques. They can be effective when combined with other approaches like rule-based systems or deep learning models.

## 3.3 Deep Learning Techniques

Deep learning [6] models have elevated hate speech detection [21], outperforming traditional machine learning capabilities. This paradigm shift is driven by the following factors.

- The unique ability of deep learning models to automatically learn intricate features and representations directly from raw text data. Unlike traditional models reliant on manual feature engineering, deep learning models leverage multiple layers of neural networks to learn complex patterns and representations hierarchically [22]. This enables them to capture underlying semantics, contextual information, and sequential dependencies in hate speech, which are often challenging to capture using traditional feature engineering approaches. This results in improved accuracy and robustness in hate speech detection.
- Furthermore, deep learning models are proficient in handling large datasets [22]. They thrive on vast amounts of labelled hate speech data, allowing them to generalize well to unseen instances. With the increasing availability of labelled datasets, deep learning models are trained on substantial data, enabling them to recognize diverse manifestations of hate speech across contexts, languages, and social media platforms. This extensive exposure helps models adapt to the ever-evolving landscape of online hate speech.

Recurrent neural networks (RNNs) [23] and their variants, such as bidirectional RNN (BRNN) [24], Gated Recurrent Units (GRU) [25], Long Short-Term Memory (LSTM) [26], and bidirectional LSTM (BiLSTM) [27], stand out for hate speech detection. LSTMs are specifically designed to capture long-range dependencies in data, whereas simple RNNs are often limited in this regard, as they tend to focus more on the immediate context and forget information from earlier parts of the input sequence [26]. Specializing in sequence modelling, they capture the sequential nature of the text and understand contextual dependencies between words. Analyzing text step by step, these models retain and update information in memory cells, enabling them to grasp context and meaning, crucial for identifying hate speech indicators.

In addition, convolutional neural networks (CNNs) [28] have also been employed in hate speech detection [29]. CNNs leverage filters or kernels to extract local features from the text, capturing patterns in different parts of the input [30]. By sliding these filters over the text and applying non-linear operations, the model can learn discriminative features related to hate speech. This allows the model to identify key phrases or combinations of words that may indicate hate speech.

Many of the above models often rely on pre-trained word embeddings such as **Word2Vec** [12], **GloVe** [31], or **MUSE** [32] for the purpose of their initialization.

## 3.4 Transformer-Based Techniques

Transformer-based models have revolutionized hate speech detection, offering significant advantages over traditional deep learning models like RNNs and LSTMs [33, 34]. Their ability to capture long-range dependencies [35], effectively handle context, and learn global relationships within text data makes them particularly well-suited for this challenging task. Some of the factors that give the transformer models their power are discussed below.

- **Long-range dependencies**: The driving force behind the adoption of transformer-based models lies in their ability to capture long-range dependencies, adeptly manage context, and comprehend global relationships within textual data. In contrast to RNNs and LSTMs, transformers employ a self-attention mechanism, enabling them to focus on various segments of the input sequence simultaneously [35]. This allows them to track dependencies between distant words and effectively model context throughout the entire text—a quality that proves particularly beneficial in addressing the complex and context-dependent patterns often exhibited in hate speech posts.
- **Pre-training and fine-tuning**: Models such as BERT [36] (Bidirectional Encoder Representations from Transformers) or RoBERTa [37] (Robustly Optimized BERT Pretraining Approach) are pre-trained on large corpora, learning rich representations of language as well as general linguistic rules. This pre-training phase enables the models to capture general contextual information, making them more robust and adaptable when fine-tuned for specific hate speech detection tasks [36, 37].

- **Multilingual capability**: Another motivation for using transformer-based models is their ability to handle diverse and multilingual (mBERT [36], XLM-RoBERTa [37]) text data. Hate speech manifests in various languages and cultural contexts, requiring models that can effectively process and understand text in multiple languages. Transformers excel in this regard as they can be trained on multilingual datasets, allowing them to capture the nuances and patterns of hate speech transcending linguistic boundaries. This multilingual capability is crucial for platforms with a global user base, where hate speech detection must account for various languages and cultural contexts.
- **Subword tokenization**: Transformer-based models leverage subword tokenization techniques such as Byte-Pair Encoding (BPE) [38], WordPiece [39], or SentencePiece [40], enabling them to handle out-of-vocabulary (OOV) words and rare or unseen terms. By breaking down words into subword units, transformers can effectively represent and understand the structure of complex words or slang often used in hate speech. This subword tokenization technique further enhances the models' ability to capture hate speech indicators that may involve variations or modifications of common words.

## 3.5 Hybrid Techniques

The modelling techniques we have covered so far are primarily generalized classification techniques. Their applicability extends beyond hate speech detection to a wide range of classification tasks. Now, let us explore some of the hybrid approaches being developed by researchers for detecting hateful content.

### 3.5.1 HurtBERT

While we previously explored the use of hate lexicons in detecting hateful content, they do have inherent limitations. Over time, transformer-based models like BERT, RoBERTa, and others have emerged as state-of-the-art solutions. Koufakou et al. [41] demonstrated how a meaningful combination of lexical features derived from a hate lexicon called HurtLex [42], into the BERT model can significantly enhance the performance of hate speech and abusive speech detection. HurtLex is a multilingual lexicon that contains words and phrases that are commonly used in abusive language. *HurtBERT* has two variants: *HurtBERT-Enc* and *HurtBERT-Emb*.

Both variants utilize two inputs: (a) sentence tokens (BERT's standard input) and (b) a vector created based on the categories found in HurtLex within the data. In both variants, the text corpus undergoes initial processing through BERT layers and is subsequently passed to another dense layer. However, the extraction of HurtLex features differs between the two variations.

## 3.5 Hybrid Techniques

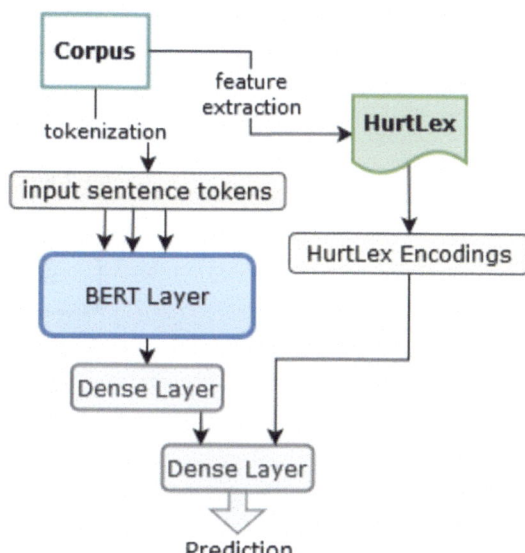

**Fig. 3.2** HurtBERT-Enc [41]

In *HurtBERT-Enc* (refer to Fig. 3.2), HurtLex categories are identified based on words in the training set, resulting in a vector known as HurtLex encoding. HurtLex consists of 17 categories, resulting in a 17-dimensional encoding. Each element in this vector represents a frequency count for the corresponding category in HurtLex. For example, if a training instance (e.g., a tweet) contains three words categorized as ethnic slurs in HurtLex, the corresponding entries in the HurtLex encoding are set to one.

In the case of *HurtBERT-Emb* (refer to Fig. 3.3), instead of creating a single vector of category-wise frequency counts, HurtLex embeddings are obtained for each word separately. To start with, HurtLex embedding is initialized as a 17-dimensional one-hot encoding representing the presence of a word in each of the lexicon categories. This is then processed through an LSTM and a dense layer. One key difference between *HurtBERT-Enc* and *HurtBERT-Emb* is that encoding is done on a comment/post level while embedding is done at a word level. Therefore, for HurtLex encodings, each record (e.g., a tweet) generates a single 17-dimensional vector, referred to as HurtLex encoding. On the other hand, for HurtLex embeddings, every word in a comment has its own 17-dimensional vector representation, known as HurtLex embedding. In both variants, the dense layer output from the BERT and the dense layer output from the HurtLex are concatenated and passed into another dense layer, which serves as the prediction layer with a sigmoid activation function.

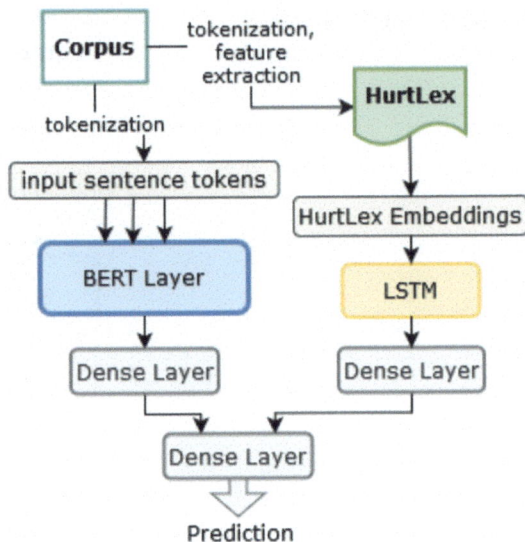

**Fig. 3.3** HurtBERT-Emb [41]

### 3.5.2 HateBERT

HateBERT [43] is a specialized model designed for the detection of abusive language in English. Built upon the BERT base-uncased[1] architecture, known for its effectiveness in various natural language processing tasks, HateBERT has been fine-tuned using **RAL-E**, a substantial dataset of English Reddit comments from communities that were banned due to their offensive, abusive, or hateful content. This extensive dataset comprises over 10 million comments. It is meticulously labelled to cover a range of abusive language phenomena, including hate speech, offensive language, and insulting expressions.

One limitation of existing pre-trained models lies in their broad language variety, which makes them suitable for general-purpose language understanding but restricts their performance with more domain-specific language nuances. The concept of re-training BERT arises from studies demonstrating the effectiveness of utilizing messages from potentially abuse-oriented online communities to generate hate embeddings or biased word embeddings. Following this premise, the approach here leverages biased embeddings created using messages from Reddit's banned communities, emphasizing a targeted focus on abusive language and hate speech detection.

---

[1] https://huggingface.co/bert-base-uncased.

### 3.5.3 HateNet

HateNet [44] is a graph convolutional network (GCN) [45] based model that learns to capture the relationships between tweets in order to identify hate speech better. HateNet addresses several challenges in hate speech detection, including the following (see Fig. 3.4).

- **Class imbalance**: Hate speech datasets are often imbalanced, with a small number of hateful tweets and a large number of non-hateful tweets. HateNet addresses this challenge by using a data augmentation technique called SubDQE to generate new hateful tweets.
- **Sparsity**: Hateful tweets often contain sparse information, making it difficult for traditional machine-learning models to detect. HateNet addresses this challenge by using a GCN, which can learn to capture the relationships between tweets in order to identify hate speech better.
- **Noisy network language**: Social media data is often noisy, containing misspellings, slang, and emojis. HateNet addresses this challenge by using pre-trained language models, like BERT, to learn noise-robust word embeddings.

HateNet consists of four core components: short text data augmentation, a semantic similarity graph, graph convolution, and weighted DropEdge. The model takes a set of tweets as input, initially undergoing a novel short text data augmentation (SubDQE). This process involves dynamic query expansion (DQE) on the dataset to identify expanded queries (parts 1 and 2 of Fig. 3.5). Subsequently, a substitution technique is applied to thew representative data using synonym replacement and word embedding vector proximity to generate new data (part 3 of Fig. 3.5). The augmented dataset is then used to construct a fully connected graph, with word/sentence embeddings as nodes and semantic cosine similarities as edges. Textual graph convolution is applied to predict whether a tweet is hateful (see Fig. 3.4). During training, the weighted DropEdge technique selectively removes edges from the input graph based on probabilities determined by semantic similarity scores.

## 3.6 Evaluation of Hate Speech Detection Systems

Having explored various hate speech detection techniques, we now direct our attention to a critical aspect: evaluation. While these techniques undeniably represent significant advancements in the fight against online hate, the journey toward accurate and effective hate speech detection extends beyond their development. It leads us to the pivotal phase of evaluation. As our focus shifts from technique development, we find ourselves at the juncture where real-world effectiveness is put to the test. Also, the potential for bias in hate speech detection systems should not be disregarded. The data used to train these systems can inadvertently introduce biases, leading to the misidentification or under-recognition of hate speech directed at certain groups or individuals. This highlights the crucial role of

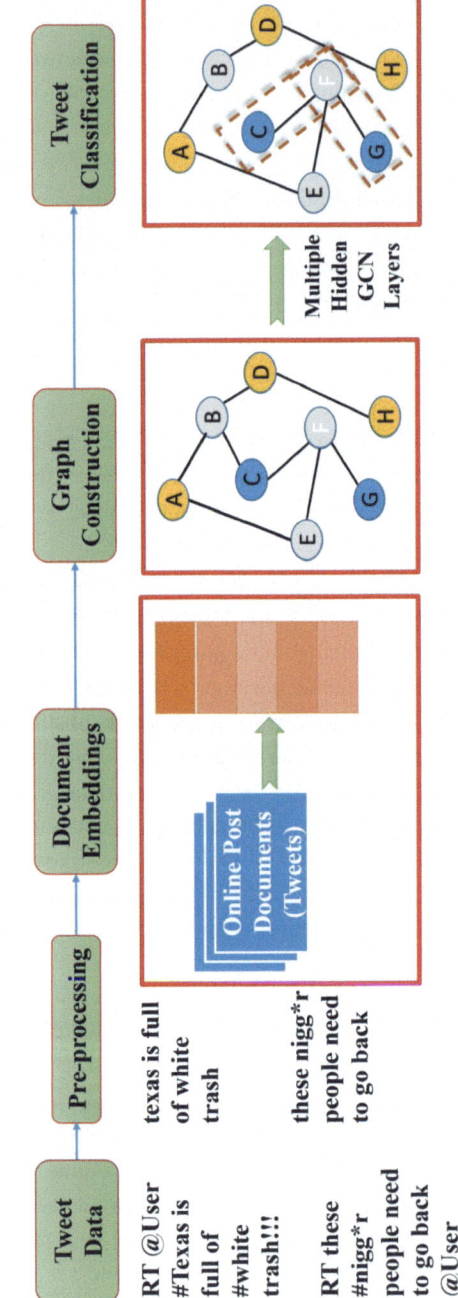

**Fig. 3.4** Overview of the HateNet framework for hate speech detection [44]

## 3.6 Evaluation of Hate Speech Detection Systems

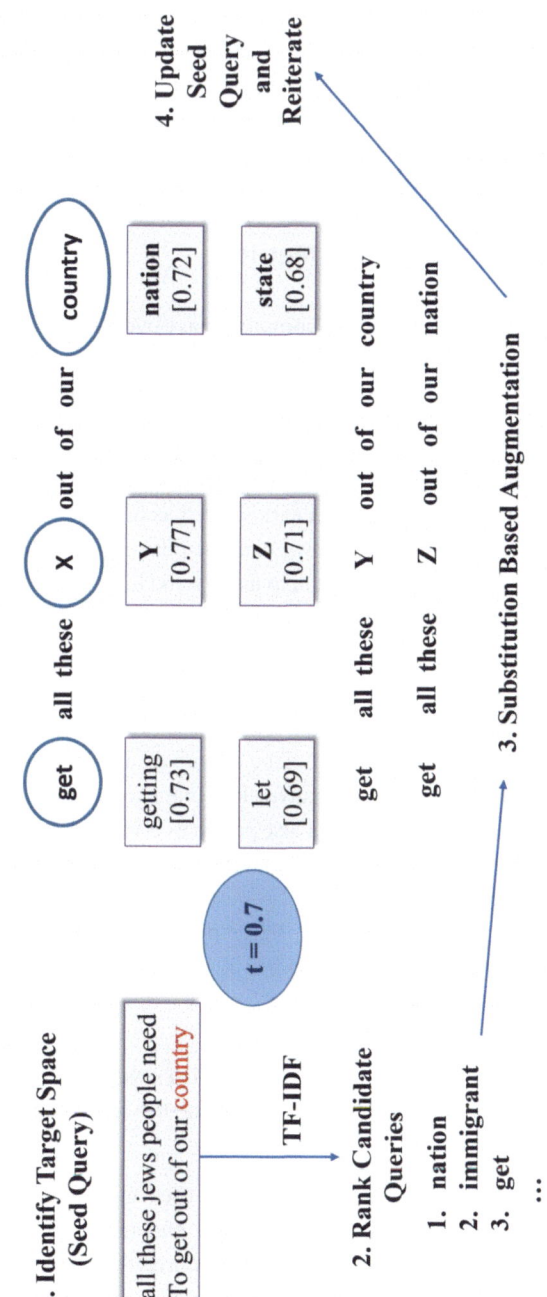

**Fig. 3.5** Overview of the SubDQE Process for Short Text Data Augmentation. Perturbed copies of the original short text are generated based on high cosine similarity and POS-tag matching after queries have been flagged. In Part 3, please note that even though 'getting' exceeds the similarity threshold, it is a gerund, while 'get' is a regular verb [44]

evaluating these systems for fairness and inclusivity. In the sections that follow, we will delve into a spectrum of evaluation metrics and strategies, shedding light on their strengths and limitations. Our goal is to provide a comprehensive overview of the evaluation landscape for hate speech detection systems, empowering researchers and practitioners with robust and reliable frameworks.

### 3.6.1 Evaluation Metrics

This section provides a comprehensive overview of the most common evaluation metrics [46] for hate speech detection. We consider hate speech detection as a binary classification task to contextualize the different metrics. Further, we will address the adjustments required for multi-class classification.

**Confusion matrix for binary classification**: Confusion matrices are commonly used in binary classification scenarios, such as hate speech detection, where the goal is to classify instances into one of two classes: positive (e.g., hate speech) or negative (e.g., non-hate speech). These metrics help understand how a classification model performs by quantifying the different classification outcomes.

- *True positive* (TP): True positives are the cases where the model correctly predicts instances of the positive class (e.g., correctly identifying hate speech).
- *True negative* (TN): True negatives are the cases where the model correctly predicts instances of the negative class (e.g., correctly identifying non-hate speech).
- *False positive* (FP): False positives occur when the model incorrectly predicts an instance as positive when it is actually negative (e.g., mistakenly identifying non-hate speech as hate speech).
- *False negative* (FN): False negatives occur when the model incorrectly predicts an instance as negative when it is actually positive (e.g., missing hate speech in a document).

Confusion matrices help us derive several important evaluation metrics, including accuracy, precision, recall, F1-Score, and the area under the ROC curve (AUC-ROC), which we discuss below.

**Accuracy**: Accuracy is a widely used metric for hate speech detection. It quantifies the percentage of correct predictions, encompassing both true positives (correctly identified hate speech) and true negatives (correctly identified non-hate speech), in relation to the total number of predictions. Accuracy is a relatively straightforward metric to calculate and interpret. This is mathematically expressed in Eq. 3.1.

$$\text{Accuracy} = \frac{\text{TP} + \text{TN}}{\text{TP} + \text{FP} + \text{TN} + \text{FN}} \qquad (3.1)$$

## 3.6 Evaluation of Hate Speech Detection Systems

However, accuracy can be misleading in certain contexts. For example, if the dataset contains a large number of non-hate speech instances, a simple system that always predicts non-hate speech can achieve high accuracy, even though it does not detect any hate speech. To address this limitation, accuracy is often used in conjunction with other metrics, such as precision, recall, and F1-score. These metrics provide a more nuanced understanding of the system's performance, considering both false positives and false negatives.

**Precision**: Precision, often referred to as positive predictive value, is a fundamental metric in the assessment of hate speech detection systems. It measures the proportion of true positive predictions (correctly identified hate speech) relative to all positive predictions made by the system. Mathematically, precision is expressed in Eq. 3.2.

$$\text{Precision} = \frac{TP}{TP + FP} \quad (3.2)$$

High precision implies that the system is effective in minimizing false alarms, ensuring that a substantial majority of the identified hate speech instances are indeed valid. However, overly emphasizing precision may lead to the system missing true hate speech instances, resulting in lower coverage.

**Recall**: Recall, also known as sensitivity or true positive rate, is the metric that quantifies the ability of a system to capture all instances of hate speech present in the dataset. This is mathematically expressed in Eq. 3.3.

$$\text{Recall} = \frac{TP}{TP + FN} \quad (3.3)$$

A high recall value indicates that the system successfully identifies most of the actual hate speech instances, but it may come at the cost of an increased number of false positives. Achieving a balance between precision and recall is often a challenging trade-off.

**F1-Score**: The F1-Score is the harmonic mean of precision and recall and offers a balanced assessment of the system's performance. It is expressed as in Eq. 3.4.

$$\text{F1-Score} = 2 \times \left( \frac{\text{Precision} \times \text{Recall}}{\text{Precision} + \text{Recall}} \right) \quad (3.4)$$

The F1-Score becomes especially useful when there is a need to optimize both precision and recall simultaneously. It provides a single metric that can help evaluate the overall effectiveness of the system. This is particularly important for hate speech detection, as false positives can have significant negative consequences.

**Area under the ROC curve (AUC-ROC)**: The AUC-ROC metric assesses the system's discriminatory power across different classification thresholds, measuring its ability to distinguish between hate speech and non-hate speech instances. The ROC curve graphically illustrates the trade-off between true positive rate (recall) and false positive rate, with the AUC-ROC representing the area under this curve (refer to Fig. 3.6). An excellent model has

**Fig. 3.6** AUC—ROC curve

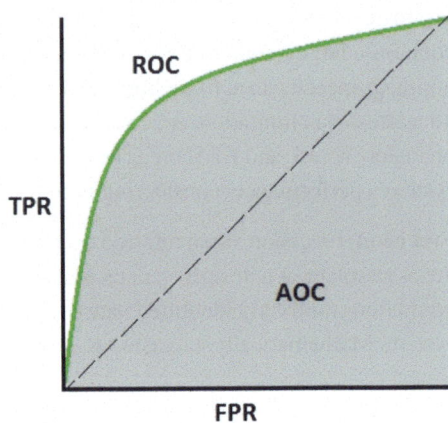

**Table 3.1** Confusion matrix for $n$ classes

| Actual classes | Predicted classes | | | |
|---|---|---|---|---|
| | Class 1 | Class 2 | ... | Class $n$ |
| Class 1 | $TP_1$ | $FP_1$ | ... | $FP_1$ |
| Class 2 | $FP_2$ | $TP_2$ | ... | $FP_2$ |
| ... | ... | ... | ... | ... |
| Class $n$ | $FP_n$ | $FP_n$ | ... | $TP_n$ |

an AUC close to 1, indicating high separability. Conversely, a poor model has an AUC near 0, suggesting the opposite, where it predicts hate speech as non-hate speech and vice versa. An AUC of 0.5 signifies that the model lacks class separation ability.

**Multi-class classification**: In multi-class classification, there are more than two classes (labels) to predict. This means that there are multiple categories into which instances can be classified (see Table 3.1). Adjustments are needed to apply the same principles to evaluate the performance of a multi-class hate speech detection system. We calculate the number of true positives, true negatives, false positives, and false negatives for each class. Each class is treated as a binary classification problem against the rest (one-vs-all), and the confusion matrices are computed accordingly. Hence if we have $n$ classes, we need to evaluate precision, recall, and F1-score for each class separately. We need a single precision score, not $n$, to compare one classifier with another. Hence, we need a way to represent precision across all classes. This is where the averaging techniques come in. There are three averaging techniques applicable to multi-class classification.

- **Macro**: A straightforward technique that calculates the arithmetic mean of metrics across all classes, affording equal weight to each class. It is most effective in situations where classification tasks are well-balanced.

## 3.6 Evaluation of Hate Speech Detection Systems

- **Weighted**: This approach takes into account the imbalance among classes. It computes the average of binary metrics while considering the number of samples in each class, thus providing a more equitable evaluation.
- **Micro**: Here we initially create confusion matrices for all classes. We then consolidate the true positives, true negatives, false positives, and false negatives across these distinct matrices, denoting them as $\Sigma TP$, $\Sigma TN$, $\Sigma FP$, and $\Sigma FN$. Subsequently, we treat these aggregated measures as if they form a single confusion matrix and employ them to calculate the desired evaluation metrics.

**Limitations and trade-offs** While these metrics offer interesting insights into the performance of hate speech detection systems, they come with limitations and trade-offs. For instance, optimizing for high precision may lead to lower recall, potentially missing significant hate speech instances. Conversely, maximizing recall may result in more false positives, which can burden human moderators with a higher workload. Therefore, the choice of which metric to prioritize depends on the specific context, goals, and requirements of the application.

### 3.6.2 HateCheck Test Suite

The evaluation metrics discussed above are commonly used to assess the performance of hate speech detection models; they have significant limitations in evaluating the nuanced and complex challenges of hate speech detection and are often insufficient to determine specific areas where the model may struggle. For instance, accuracy can indicate how many instances were correctly classified; however, it does not differentiate between the types of mistakes the model makes—whether it tends to misclassify counter-speech as hate speech or fails to detect hate directed at less frequently targeted groups. Similarly, precision and recall offer insight into false positives and false negatives, but they do not explain *why these errors occur* or how the model handles different forms of hate speech, such as slurs, veiled language, or hate expressed through comparisons or sarcasm. These general metrics focus on aggregate performance, potentially masking essential deficiencies in the model's ability to handle specific linguistic patterns, diverse speech forms, or subtle manipulations of language.

The **HateCheck Test Suite** [47] significantly enhances the evaluation of hate speech detection models. Unlike traditional metrics that measure overall performance, HateCheck provides a more precise assessment comprising **29 functional tests** divided into 11 categories. These tests systematically examine different aspects of hate speech detection, with **18 tests dedicated** to various expressions of **hateful** content and **11 tests focused on non-hateful** content. HateCheck evaluates how well a model handles slurs against different target groups, how it deals with implicit derogation, whether it can differentiate between hate speech and counter-speech (e.g., someone defending a marginalized group), and many more. By focusing on these distinctive linguistic phenomena, HateCheck divulges whether a model is robust across various contexts rather than just performing well in aggregate.

The critical advantage of HateCheck is its **granular evaluation**. It helps researchers and practitioners pinpoint precisely where a model succeeds or fails, offering a detailed view of its strengths and weaknesses. For example, a model might perform well in detecting direct hate speech but struggle with indirect hate that uses subtle language or sarcasm. Such weaknesses might not be captured in traditional evaluation metrics, where high accuracy or F1 scores give a false sense of reliability. By contrast, HateCheck's tests make it possible to identify these gaps in the model's performance, providing an opportunity for targeted improvements.

Likewise, HateCheck is valuable for ensuring **fairness and reliability** across diverse groups. Many hate speech detection models exhibit bias, over-identifying hate speech against certain groups or failing to detect hate directed at others. HateCheck addresses this issue by including tests that involve various target groups, ensuring that a model's performance is balanced and not skewed towards a particular type of speech or group. This is crucial for developing fair and trustworthy models in real-world applications, where failing to detect hate speech against specific groups or disproportionately flagging benign speech from certain demographics can have serious consequences.

## 3.7 A Review of Key Results

In this section, we shall review some of the key results obtained for various datasets across the machine learning and deep learning models as well as the hybrid models.

### 3.7.1 Results from Machine and Deep Learning Models

Table 3.2 presents the performance of various machine learning models in terms of macro F1-score for hate speech detection across multiple languages. These approaches include traditional models, deep learning models, and transformer-based models [48]. The authors also examine the impact of varying training sizes, particularly in low-resource settings. As expected, performance increases with increasing training data. However, the relative performance of different models varies depending on the language. Several key observations can be made.

- First, **LASER + LR** consistently exhibits the best performance in low-resource settings (16, 32, 64, 128, 256) for all languages.
- Second, **MUSE + CNN-GRU** performs the worst in almost all scenarios.

## 3.7 A Review of Key Results

**Table 3.2** The performance of some popular models across several languages. Here, Full D represents the full training data. The **bold** figures represent the best scores, and <u>underline</u> represents the second best. [48]

| Language | Model | Training size | | | | | |
|---|---|---|---|---|---|---|---|
| | | 16 | 32 | 64 | 128 | 256 | Full D |
| Arabic | MUSE + CNN-GRU | 0.4412 | 0.4438 | 0.4486 | 0.4664 | 0.5818 | 0.7368 |
| | Translation + BERT | 0.4555 | 0.4495 | <u>0.5551</u> | 0.5448 | 0.7017 | <u>0.8115</u> |
| | LASER + LR | **0.5533** | **0.6755** | **0.7304** | **0.7488** | **0.7698** | 0.7920 |
| | mBert | <u>0.4588</u> | <u>0.4533</u> | 0.4408 | <u>0.6486</u> | <u>0.7295</u> | **0.8320** |
| English | MUSE + CNN-GRU | <u>0.4580</u> | <u>0.4594</u> | <u>0.4653</u> | 0.4646 | <u>0.4813</u> | 0.6441 |
| | BERT | 0.4071 | 0.3925 | 0.4260 | <u>0.4720</u> | 0.4578 | **0.7143** |
| | LASER + LR | **0.4617** | **0.4899** | **0.5376** | **0.5624** | **0.5885** | 0.6526 |
| | mBert | 0.1773 | 0.3251 | 0.4488 | 0.4578 | 0.4578 | <u>0.7101</u> |
| German | MUSE + CNN-GRU | 0.4708 | 0.4708 | 0.4708 | 0.4708 | 0.4762 | 0.5756 |
| | Translation + BERT | 0.4812 | <u>0.4758</u> | <u>0.4719</u> | <u>0.4729</u> | 0.4724 | **0.7662** |
| | LASER + LR | <u>0.4974</u> | **0.5201** | **0.5465** | **0.5925** | **0.6488** | <u>0.6873</u> |
| | mBert | **0.5037** | 0.4750 | 0.4708 | 0.4717 | <u>0.5022</u> | 0.6517 |
| Indonesian | MUSE + CNN-GRU | 0.4250 | 0.4823 | 0.5263 | 0.5354 | 0.5890 | 0.7110 |
| | Translation + BERT | 0.4957 | 0.5003 | 0.5179 | 0.5682 | 0.6341 | 0.7670 |
| | LASER + LR | **0.5226** | **0.5376** | **0.5882** | **0.6259** | **0.6890** | <u>0.7872</u> |
| | mBert | <u>0.5106</u> | <u>0.5219</u> | <u>0.5414</u> | <u>0.6016</u> | <u>0.6530</u> | **0.8119** |
| Italian | MUSE + CNN-GRU | 0.4055 | 0.4476 | 0.4461 | 0.5206 | 0.5965 | 0.7349 |
| | Translation + BERT | 0.5006 | <u>0.5943</u> | <u>0.6215</u> | <u>0.6678</u> | 0.6919 | 0.7922 |
| | LASER + LR | <u>0.5688</u> | **0.6210** | **0.6843** | **0.7175** | **0.7347** | <u>0.7996</u> |
| | mBert | **0.5774** | 0.4567 | 0.5834 | 0.6664 | <u>0.7026</u> | **0.8260** |

**Table 3.2** (continued)

| Language | Model | Training size | | | | | |
|---|---|---|---|---|---|---|---|
| | | 16 | 32 | 64 | 128 | 256 | Full D |
| Polish | MUSE + CNN-GRU | 0.4842 | 0.4842 | 0.4841 | 0.4842 | 0.5180 | 0.6337 |
| | Translation + BERT | 0.4842 | 0.4853 | 0.4842 | 0.4842 | 0.5066 | **0.7161** |
| | LASER + LR | **0.4889** | **0.4879** | **0.5360** | **0.5739** | **0.6172** | 0.6439 |
| | mBert | 0.4829 | 0.4847 | 0.4842 | 0.4842 | 0.4842 | 0.7069 |
| Portuguese | MUSE + CNN-GRU | 0.4480 | 0.3807 | 0.4184 | 0.4228 | 0.4562 | 0.6100 |
| | Translation + BERT | 0.4532 | 0.4893 | 0.4712 | 0.5102 | 0.5994 | 0.6935 |
| | LASER + LR | **0.5194** | **0.5536** | **0.6070** | **0.6210** | **0.6412** | **0.6941** |
| | mBert | 0.5154 | 0.4245 | 0.4148 | 0.5493 | 0.5745 | 0.6713 |
| Spanish | MUSE + CNN-GRU | 0.4382 | 0.3354 | 0.3558 | 0.4203 | 0.4995 | 0.6364 |
| | Translation + BERT | 0.4598 | 0.4722 | 0.5080 | 0.4576 | 0.6035 | 0.7237 |
| | LASER + LR | **0.5168** | **0.5434** | **0.5521** | **0.5938** | **0.6153** | 0.6997 |
| | mBert | 0.4395 | 0.4285 | 0.4048 | 0.4861 | 0.5999 | **0.7329** |
| French | MUSE + CNN-GRU | 0.4878 | 0.4683 | 0.5008 | 0.5222 | 0.5250 | 0.5619 |
| | Translation + BERT | 0.4173 | 0.4260 | 0.4429 | 0.4749 | 0.6037 | **0.6595** |
| | LASER + LR | **0.5058** | **0.5486** | **0.6136** | **0.6302** | **0.6085** | 0.6172 |
| | mBert | 0.4818 | 0.4139 | 0.4053 | 0.4355 | 0.5701 | 0.6165 |

- Third, **Translation + BERT**[2] demonstrates competitive performance, particularly in languages such as German, Polish, Portuguese, and Spanish.

Overall, there is no single "one size fits all" solution for hate speech detection across languages. **Translation + BERT**, however, emerges as a promising candidate due to its consistent performance. Further improvement in translation quality for certain languages could potentially enhance the effectiveness of this model. While **LASER + LR** performs well in low-resource settings, BERT-based models like **Translation + BERT** (English, German, Polish, and French) and **mBERT** (Arabic, Indonesian, Italian, and Spanish) outperform it

---

[2] The input sentence is first translated to the English language, which is then provided as input to the BERT model.

when sufficient data is available. Interestingly, even with only 256 training data points, some BERT-based models, such as **Translation + BERT** (Spanish, French) and **mBERT** (Arabic, Indonesian, Italian), achieve performance close to **LASER + LR**.

### 3.7.2 Performance of HurtBERT

Koufakou et al. [41] conducted comprehensive experiments across six datasets to showcase the effectiveness of their proposed model. Using BERT as a baseline, they evaluated performance in both in-domain (same dataset for training and testing) and out-of-domain (training on one dataset, testing on another) settings. The results are summarized in Table 3.3, using macro F1-score as the metric. HurtBERT consistently outperforms the baseline model across four of the six datasets—-AbuseEval [49], HatEval [50], OLID [51], and Waseem [13]. Notably, *HurtBERT-Emb* consistently exhibits superior performance in all four cases. As expected, out-of-domain results are lower than in-domain results. For instance, in the Davidson dataset, in-domain performance consistently exceeds 90%, while out-of-domain performance (training on other datasets and testing on Davidson) varied from 40% to 70%. In cross-dataset experiments, both variants of HurtBERT consistently outperform the baseline, demonstrating their superior ability to generalize to new data. In summary, the integration of lexical knowledge from HurtLex into the BERT model significantly improves the efficacy of abusive and hate speech detection.

### 3.7.3 Performance of HateNet

The performance of the HateNet [44] model was evaluated using various embeddings, including bag of words (BOW), TF-IDF, Word2Vec, GloVe, FastText, BERT [36], DistilBERT (DBERT) [53], and SentenceBERT (SBERT) [54]. In order to demonstrate HateNet's effectiveness, six other traditional machine-learning classifiers were also analyzed as a baseline. The performance of all models has been summarized in Table 3.5, utilizing the dataset created by Davidson et al. [2]; notably, sentence BERT + HateNet outperforms all other methods.

### 3.7.4 Performance of HateBERT

To assess the effectiveness of HateBERT in identifying abusive language, a series of experiments [43] were carried out on three English datasets—OffensEval 2019 [51], AbuseEval [49], and HatEval [50]. These datasets exhibit class imbalances between positive and negative labels and cover a range of language phenomena that vary in specificity. This diversity permitted the authors to evaluate both the robustness and portability of HateBERT. Table 3.4

**Table 3.3** Performance across datasets in terms of macro-F1 score. Shaded rows represent in-domain performance, and other rows represent out-domain performance. B stands for the baseline BERT. HB-Enc stands for HurtBERT-Enc, and HB-Emb stands for HurtBERT-Emb. Bold indicates HurtBERT improves on the baseline; underlined indicates the best result (max) [41]

| Train set | AbuseEval | | | Davidson | | | Founta | | |
|---|---|---|---|---|---|---|---|---|---|
| Test Set | B | HB-Enc | HB-Emb | B | HB-Enc | HB-Emb | B | HB-Enc | HB-Emb |
| AbuseEval [49] | 0.659 | **0.669** | **0.686** | 0.577 | **0.578** | **0.583** | 0.672 | 0.657 | 0.671 |
| Davidson [2] | 0.462 | 0.444 | 0.453 | 0.908 | 0.907 | 0.907 | 0.742 | 0.738 | **0.745** |
| Founta [52] | 0.707 | **0.715** | 0.702 | 0.849 | **0.850** | **0.850** | 0.916 | 0.914 | 0.913 |
| HatEval [50] | 0.579 | 0.579 | 0.571 | 0.515 | **0.519** | **0.517** | 0.532 | **0.539** | **0.541** |
| HatEval Mig [50] | 0.569 | 0.554 | 0.559 | 0.533 | **0.542** | **0.546** | 0.542 | **0.544** | **0.578** |
| HatEval Mis [50] | 0.572 | **0.582** | 0.567 | 0.307 | **0.308** | 0.306 | 0.341 | **0.355** | **0.348** |
| OLID [51] | 0.638 | **0.662** | **0.666** | 0.663 | **0.667** | **0.674** | 0.753 | 0.741 | 0.753 |
| Waseem [13] | 0.589 | **0.596** | 0.583 | 0.629 | **0.636** | **0.636** | 0.602 | 0.600 | **0.612** |
| Train set | HatEval | | | OLID | | | Waseem | | |
| Test set | B | HB-Enc | HB-Emb | B | HB-Enc | HB-Emb | B | HB-Enc | HB-Emb |
| AbuseEval [49] | 0.562 | 0.548 | 0.552 | 0.663 | **0.666** | **0.680** | 0.521 | 0.520 | **0.541** |
| Davidson [2] | 0.583 | 0.547 | 0.551 | 0.703 | **0.704** | 0.703 | 0.406 | **0.445** | **0.462** |
| Founta [52] | 0.570 | 0.543 | 0.554 | 0.874 | **0.877** | 0.874 | 0.512 | **0.516** | **0.540** |
| HatEval [50] | 0.533 | **0.553** | **0.562** | 0.535 | **0.537** | **0.540** | 0.524 | 0.524 | **0.542** |
| HatEval Mig [50] | 0.463 | **0.486** | **0.483** | 0.575 | 0.549 | **0.578** | 0.420 | **0.436** | **0.450** |
| HatEval Mis [50] | 0.598 | **0.638** | **0.633** | 0.361 | **0.376** | **0.371** | 0.588 | 0.579 | **0.595** |
| OLID [51] | 0.565 | 0.545 | 0.549 | 0.739 | 0.739 | **0.747** | 0.511 | 0.507 | **0.536** |
| Waseem [13] | 0.632 | 0.614 | 0.620 | 0.632 | 0.610 | **0.637** | 0.836 | 0.834 | **0.838** |

summarizes the performance across all datasets. It is evident that HateBERT significantly outperforms the generic BERT model in these experiments (Table 3.4).

**Takeaways**: Overall, transformer-based models have emerged as a significant advancement in hate speech detection [55, 56] due to their ability to capture long-range dependencies, leverage pre-training and fine-tuning techniques, handle diverse languages, and effectively represent complex words and phrases. By harnessing these capabilities, transformer-based models can provide more accurate, context-aware, and adaptable hate speech detection solutions, contributing to the creation of safer and more inclusive online environments.

## 3.8 Multimodal Information

**Table 3.4** Performance across datasets for BERT and HateBERT. The best score between both is represented in bold

| Dataset | Model | Macro F1 | Pos. class–F1 |
|---|---|---|---|
| OffensEval 2019 [51] | BERT | 0.803±.006 | 0.715±.009 |
|  | HateBERT | **0.809±0.008** | **0.723±0.012** |
| AbusEval [49] | BERT | 0.727±.008 | 0.552±.012 |
|  | HateBERT | **0.765±0.006** | **0.623±0.010** |
| HatEval [50] | BERT | 0.480±.008 | 0.633±.002 |
|  | HateBERT | **0.516±0.007** | **0.645±0.001** |

**Table 3.5** Macro-F1 Score of HateNet and Baseline models. Bold represents the best performance [44]

|  | LR | NB | KNN | SVM | XGB | MLP | HateNet |
|---|---|---|---|---|---|---|---|
| SBERT | 0.810 | 0.731 | 0.702 | 0.832 | 0.821 | 0.824 | **0.843** |
| BERT | 0.766 | 0.597 | 0.624 | 0.772 | 0.762 | 0.787 | 0.791 |
| DBERT | 0.807 | 0.657 | 0.703 | 0.812 | 0.806 | 0.831 | 0.816 |
| GloVe | 0.796 | 0.623 | 0.648 | 0.819 | 0.814 | 0.803 | 0.811 |
| W2V | 0.759 | 0.634 | 0.639 | 0.816 | 0.817 | 0.808 | 0.789 |
| FastText | 0.755 | 0.543 | 0.604 | 0.796 | 0.806 | 0.777 | 0.776 |
| TF-IDF | 0.762 | 0.725 | 0.152 | 0.573 | 0.826 | 0.665 | 0.648 |
| BOW | 0.763 | 0.703 | 0.331 | 0.572 | 0.828 | 0.694 | 0.685 |

## 3.8 Multimodal Information

In Sect. 1.5, we pointed out various modalities of hate speech. While our earlier discussions primarily centred around mechanisms for detecting text-based hate speech, let us now discuss methods for identifying other forms of hate speech.

### 3.8.1 Image-Based Hate Speech Detection

Various models, including ResNet-152 [57], VGG16, VGG19 [58], DenseNet-161 [59], ResNeXt-101, Vision Transformer (VIT) [60], Visual Attention Network (VAN) [61], etc. have bee used to detect image-based hate speech. The general workflow involves utilizing these pre-trained models to extract features, which are then fed into a classification model, which can be either a traditional machine learning model (such as LR [3] or SVM [18]) or a sophisticated deep learning [6] architecture. It is important to note that depending on the chosen pre-trained model, image resizing may be necessary to pass the images via these models.

### 3.8.2 Audio-Based Hate Speech Detection

Several techniques have been used recently to detect whether an audio is hateful. One approach involves extracting the transcript from the audio using a speech recognition tool [62] and subsequently applying a text-based hate speech classification model for the detection. However, a drawback of this method lies in the potential loss of crucial elements such as tone, shouting, loudness, or emotional nuances, which may be essential to deciding whether an audio is hateful.

Another viable solution is extracting features directly from the audio signals for building classification models. Mel-Frequency Cepstral Coefficients (MFCCs) [63], capturing the spectral envelope of audio signals [64], stand out as popular feature sets widely utilized for audio classification, including detecting hate speech. In addition, spectrograms [65]—visual representations of frequencies over time—serve as effective feature extraction tools. Once the visual representation is obtained as an image, pre-trained models designed for image-based tasks can be leveraged to extract relevant features. With the rapid evolution of transformer-based models, approaches such as UniSpeech [66], Wav2Vec [67], WavLM [68], and others are gaining popularity over traditional approaches for audio processing tasks.

### 3.8.3 Hate Meme Detection

A meme typically comprises two components: a textual element and an image. Whether a meme is deemed hateful can hinge on the textual content independently, the image alone, or a synergistic combination of both [69]. While text-based and image-based models can serve as effective baselines, researchers are increasingly directing their endeavors toward developing multimodal models that intelligently integrate textual and visual features to harness the advantages of each.

One straightforward technique for leveraging both modalities is *fusion*. Fusion can be executed in various ways, with the two popular choices being (1) *concatenation* and (2) *late fusion*. In concatenation, pre-trained features from text-based and image-based models are concatenated and passed through a multilayer perceptron (MLP) for classification. In the case of late fusion, the extracted text and image features undergo MLP processing independently. The intermediate layers' features are then concatenated, and a final MLP is employed for classification [70]. In addition to these fusion techniques, several multimodal pre-trained models, such as ViLBERT [71], Visual BERT [72], UNITER [73], CLIP [74], and dual-stream models like LXMERT [75] are being actively utilized to detect hateful memes.

### 3.8.4 Hate Video Detection

A video presents a more intricate challenge compared to other forms of hate speech. It can be silent, with potentially hateful text displayed, or it may contain benign images accompanied by hateful audio. The perception of hate in a video can result from a combination of both audio and visual components. In addition, a video may incorporate still images within its frames, conveying hateful content. Given the diverse modalities associated with videos, the multitude of possible combinations is large. Consequently, detecting hate videos is inherently complex; so far, there has been a limited number of studies addressing this nuanced task.

Similar to audio-based hate speech detection, one straightforward approach involves extracting the transcript from the video and utilizing it to construct a hate video classification model. Also, audio-based features can be extracted from the video using the modelling techniques discussed in Sect. 3.8.2. Further, a video is a collection of frames (or images), and the relevant ordering of frames matters in deciding the context of a video. Hence, we can extract frames from the videos and pass these frames through a pre-trained image-based model to obtain relevant features. These features can then be passed through an LSTM network [26] to capture the sequential ordering of frames [76]. It is worth noting that an LSTM model is not the only alternative; other similar models can also be considered. In addition, akin to hate meme detection, these modalities can be fused to harness all-round information to improve performance. In Fig. 3.7, we illustrate a pipeline for hate video classification.

## 3.9 LLMs for Hate Speech Detection

So far, we have explored the development of classification models for detecting hate speech. Using these models enable platforms to rapidly address hateful content, taking actions such as removal, user warnings, or providing resources to mitigate its impact. Traditionally, a prevalent method for building such classifiers relies on labelled data. However, the process of annotating content poses significant challenges. Acquiring well-trained annotators who comprehend the task well is the initial hurdle. Multi-linguality of users on social media platforms compounds the difficulty. Moreover, the demographics of annotators may introduce biases, impacting the objectivity of annotation tasks due to variations in their social and cultural backgrounds [77].

Consequently, there is a growing need for alternative approaches to detect hate speech accurately. Recently, generative AI models like ChatGPT [78], Google Bard [79], GPT-4 [80], FLAN-T5 [81], LLaMA [82], OpenFlamingo [83], and InstructBLIP [84] have gained significant attention due to their remarkable capabilities. These models, often called *large language models* (LLMs), are trained on massive datasets and possess the ability to comprehend and generate text resembling human speech. Moreover, some LLMs can handle text and multimodal data, opening up further possibilities. The success of LLMs in hate speech

**Fig. 3.7** A schematic representation of a multi-modal hate video detection model [76]

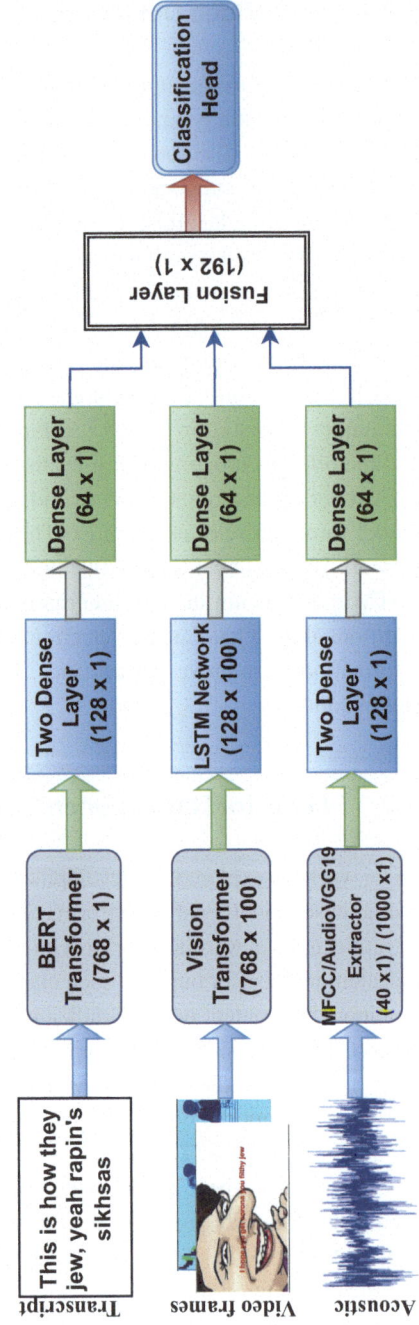

## 3.9 LLMs for Hate Speech Detection

detection hinges on two key factors: (1) the quality of their pre-training data and (2) the design of the prompts used alongside the content for classification.

### 3.9.1 Prompt Design

Prompt design refers to carefully crafting input instructions or queries given to a language model to elicit desired responses. A prompt guides the model on how to generate or classify text based on the input it receives. The design of the prompt is crucial because it influences the model's understanding of the task and can impact the quality of its output [77]. A well-designed prompt can help the model deliver more relevant and coherent responses, while a poorly designed prompt can lead to irrelevant or unsatisfactory outcomes.

It is crucial to emphasize that there is no one-size-fits-all prompt suitable for our specific purpose. What works effectively for one LLM might not yield the same results for another. The crafting of prompts is an iterative process, requiring precision in the choice of words, nuanced adjustments to tone, and, at times, a willingness to revisit the drawing board. Each iteration represents a step closer to a finely tuned prompt that resonates with the AI, leading to a creative output that aligns seamlessly with the intended version.

### 3.9.2 Examples in Action

Here we illustrate the approach with an example. In our context, the objective is to classify statements as either hate speech or not. Initially, one constructs a prompt as follows: *"Can you confirm whether the given statement qualifies as hate speech? Statement: ⟨input text⟩"*. Although humans seamlessly comprehend such instructions, the prompt might remain ambiguous to LLMs. The model may not understand the output format—whether it should be a 'yes' or 'no' or simply indicate 'hate speech' or 'not-hate speech'. Given that it is a classification task, we require a deterministic, consistent response to facilitate appropriate action. In the subsequent iteration of prompt fine-tuning, one can make the following changes: *"Can you confirm whether the given statement qualifies as hate speech, adhering to the provided definition? Respond with only 'yes' or 'no'. Statement: ⟨input text⟩"*. Now, our prompt precisely communicates the expected response format to the LLM.

After using this prompt for some time, we may observe that for certain statements, LLMs struggle to provide a desired response, expressing uncertainty about the definition of hate speech. We can further refine the prompt by incorporating additional clues to address this. In essence, crafting the perfect prompt becomes the linchpin, unlocking a realm of creative possibilities that seamlessly merge human ingenuity with machine precision.

## References

1. Md Saroar Jahan and Mourad Oussalah. A systematic review of hate speech automatic detection using natural language processing. *Neurocomputing*, page 126232, 2023.
2. Thomas Davidson, Dana Warmsley, Michael Macy, and Ingmar Weber. Automated hate speech detection and the problem of offensive language. In *Proceedings of the international AAAI conference on web and social media*, volume 11, pages 512–515, 2017.
3. Raymond E Wright. *Logistic Regression.*, 1995.
4. Irina Rish et al. An empirical study of the naive bayes classifier. In *IJCAI 2001 workshop on empirical methods in artificial intelligence*, volume 3, pages 41–46, 2001.
5. Leo Breiman. Random forests. *Machine learning*, 45:5–32, 2001.
6. Yann LeCun, Yoshua Bengio, and Geoffrey Hinton. Deep learning. *nature*, 521(7553):436–444, 2015.
7. Njagi Dennis Gitari, Zhang Zuping, Hanyurwimfura Damien, and Jun Long. A lexicon-based approach for hate speech detection. *International Journal of Multimedia and Ubiquitous Engineering*, 10(4):215–230, 2015.
8. Sean MacAvaney, Hao-Ren Yao, Eugene Yang, Katina Russell, Nazli Goharian, and Ophir Frieder. Hate speech detection: Challenges and solutions. *PloS one*, 14(8):e0221152, 2019.
9. Zeerak Talat, James Thorne, and Joachim Bingel. *Bridging the Gaps: Multi Task Learning for Domain Transfer of Hate Speech Detection*, pages 29–55. Springer International Publishing, Cham, 2018. ISBN 978-3-319-78583-7.
10. Aditya Bohra, Deepanshu Vijay, Vinay Singh, Syed Sarfaraz Akhtar, and Manish Shrivastava. A dataset of Hindi-English code-mixed social media text for hate speech detection. In Malvina Nissim, Viviana Patti, Barbara Plank, and Claudia Wagner, editors, *Proceedings of the Second Workshop on Computational Modeling of People's Opinions, Personality, and Emotions in Social Media*, pages 36–41, New Orleans, Louisiana, USA, June 2018. Association for Computational Linguistics. URL https://aclanthology.org/W18-1105.
11. Rijul Magu, Kshitij Joshi, and Jiebo Luo. Detecting the hate code on social media, 2017.
12. Tomas Mikolov, Kai Chen, Greg Corrado, and Jeffrey Dean. Efficient estimation of word representations in vector space. *arXiv preprint* arXiv:1301.3781, 2013.
13. Zeerak Waseem and Dirk Hovy. Hateful symbols or hateful people? predictive features for hate speech detection on twitter. In *Proceedings of the NAACL student research workshop*, pages 88–93, 2016.
14. Stephen Akuma, Tyosar Lubem, and Isaac Terngu Adom. Comparing bag of words and tf-idf with different models for hate speech detection from live tweets. *International Journal of Information Technology*, 14(7):3629–3635, 2022.
15. Quoc Le and Tomas Mikolov. Distributed representations of sentences and documents. In *International conference on machine learning*, pages 1188–1196. PMLR, 2014.
16. Mikel Artetxe and Holger Schwenk. Massively multilingual sentence embeddings for zero-shot cross-lingual transfer and beyond. *Transactions of the Association for Computational Linguistics*, 7:597–610, 2019.
17. Sindhu Abro, Sarang Shaikh, Zahid Hussain Khand, Ali Zafar, Sajid Khan, and Ghulam Mujtaba. Automatic hate speech detection using machine learning: A comparative study. *International Journal of Advanced Computer Science and Applications*, 11(8), 2020.
18. Lipo Wang. *Support vector machines: theory and applications*, volume 177. Springer Science & Business Media, 2005.
19. Barry De Ville. Decision trees. *Wiley Interdisciplinary Reviews: Computational Statistics*, 5(6):448–455, 2013.

## References

20. Leif E Peterson. K-nearest neighbor. *Scholarpedia*, 4(2):1883, 2009.
21. Pinkesh Badjatiya, Shashank Gupta, Manish Gupta, and Vasudeva Varma. Deep learning for hate speech detection in tweets. WWW, pages 759–760, 2017.
22. Mohammad Mustafa Taye. Understanding of machine learning with deep learning: Architectures, workflow, applications and future directions. *Computers*, 12(5):91, 2023.
23. Larry R Medsker and LC Jain. Recurrent neural networks. *Design and Applications*, 5(64-67):2, 2001.
24. Mike Schuster and Kuldip K Paliwal. Bidirectional recurrent neural networks. *IEEE transactions on Signal Processing*, 45(11):2673–2681, 1997.
25. Junyoung Chung, Caglar Gulcehre, KyungHyun Cho, and Yoshua Bengio. Empirical evaluation of gated recurrent neural networks on sequence modeling. *arXiv preprint* arXiv:1412.3555, 2014.
26. Long Short-Term Memory. Long short-term memory. *Neural computation*, 9(8):1735–1780, 2010.
27. Alex Graves, Santiago Fernández, and Jürgen Schmidhuber. Bidirectional lstm networks for improved phoneme classification and recognition. In *International conference on artificial neural networks*, pages 799–804. Springer, 2005.
28. Keiron O'Shea and Ryan Nash. An introduction to convolutional neural networks. *arXiv preprint* arXiv:1511.08458, 2015.
29. Björn Gambäck and Utpal Kumar Sikdar. Using convolutional neural networks to classify hate-speech. In *Proceedings of the first workshop on abusive language online*, pages 85–90, 2017.
30. Alon Jacovi, Oren Sar Shalom, and Yoav Goldberg. Understanding convolutional neural networks for text classification. *arXiv preprint* arXiv:1809.08037, 2018.
31. Jeffrey Pennington, Richard Socher, and Christopher D Manning. Glove: Global vectors for word representation. In *Proceedings of the 2014 conference on empirical methods in natural language processing (EMNLP)*, pages 1532–1543, 2014.
32. Alexis Conneau, Guillaume Lample, Marc'Aurelio Ranzato, Ludovic Denoyer, and Hervé Jégou. Word translation without parallel data. *arXiv preprint* arXiv:1710.04087, 2017.
33. Somnath Banerjee, Maulindu Sarkar, Nancy Agrawal, Punyajoy Saha, and Mithun Das. Exploring transformer based models to identify hate speech and offensive content in english and indo-aryan languages. 2021.
34. Binny Mathew, Punyajoy Saha, Seid Muhie Yimam, Chris Biemann, Pawan Goyal, and Animesh Mukherjee. Hatexplain: A benchmark dataset for explainable hate speech detection. *arXiv preprint* arXiv:2012.10289, 2020b.
35. Ashish Vaswani, Noam Shazeer, Niki Parmar, Jakob Uszkoreit, Llion Jones, Aidan N Gomez, Łukasz Kaiser, and Illia Polosukhin. Attention is all you need. *Advances in neural information processing systems*, 30, 2017.
36. Jacob Devlin, Ming-Wei Chang, Kenton Lee, and Kristina Toutanova. Bert: Pre-training of deep bidirectional transformers for language understanding. *arXiv preprint* arXiv:1810.04805, 2018.
37. Alexis Conneau, Kartikay Khandelwal, Naman Goyal, Vishrav Chaudhary, Guillaume Wenzek, Francisco Guzmán, Edouard Grave, Myle Ott, Luke Zettlemoyer, and Veselin Stoyanov. Unsupervised cross-lingual representation learning at scale. *arXiv preprint* arXiv:1911.02116, 2019.
38. Rico Sennrich, Barry Haddow, and Alexandra Birch. Neural machine translation of rare words with subword units. In *Proceedings of the 54th Annual Meeting of the Association for Computational Linguistics (Volume 1: Long Papers)*, pages 1715–1725, 2016.
39. Mike Schuster and Kaisuke Nakajima. Japanese and korean voice search. In *2012 IEEE international conference on acoustics, speech and signal processing (ICASSP)*, pages 5149–5152. IEEE, 2012.
40. Taku Kudo and John Richardson. Sentencepiece: A simple and language independent subword tokenizer and detokenizer for neural text processing. In *Proceedings of the 2018 Conference*

*on Empirical Methods in Natural Language Processing: System Demonstrations*, pages 66–71, 2018.
41. Anna Koufakou, Endang Wahyu Pamungkas, Valerio Basile, Viviana Patti, et al. Hurtbert: Incorporating lexical features with bert for the detection of abusive language. In *Proceedings of the fourth workshop on online abuse and harms*, pages 34–43. Association for Computational Linguistics, 2020.
42. Elisa Bassignana, Valerio Basile, Viviana Patti, et al. Hurtlex: A multilingual lexicon of words to hurt. In *CEUR Workshop proceedings*, volume 2253, pages 1–6. CEUR-WS, 2018.
43. Tommaso Caselli, Valerio Basile, Jelena Mitrović, and Michael Granitzer. Hatebert: Retraining bert for abusive language detection in english. In *Proceedings of the 5th Workshop on Online Abuse and Harms (WOAH 2021)*, pages 17–25, 2021.
44. Charles Duong, Lei Zhang, and Chang-Tien Lu. Hatenet: A graph convolutional network approach to hate speech detection. In *2022 IEEE International Conference on Big Data (Big Data)*, pages 5698–5707. IEEE, 2022.
45. Si Zhang, Hanghang Tong, Jiejun Xu, and Ross Maciejewski. Graph convolutional networks: a comprehensive review. *Computational Social Networks*, 6(1):1–23, 2019a.
46. Ž Vujović et al. Classification model evaluation metrics. *International Journal of Advanced Computer Science and Applications*, 12(6):599–606, 2021.
47. Paul Röttger, Bertie Vidgen, Dong Nguyen, Zeerak Waseem, Helen Margetts, and Janet Pierrehumbert. Hatecheck: Functional tests for hate speech detection models. In *Proceedings of the 59th Annual Meeting of the Association for Computational Linguistics and the 11th International Joint Conference on Natural Language Processing (Volume 1: Long Papers)*, pages 41–58, 2021.
48. Sai Saketh Aluru, Binny Mathew, Punyajoy Saha, and Animesh Mukherjee. A deep dive into multilingual hate speech classification. In *Machine Learning and Knowledge Discovery in Databases. Applied Data Science and Demo Track: European Conference, ECML PKDD 2020, Ghent, Belgium, September 14–18, 2020, Proceedings, Part V*, pages 423–439. Springer, 2021.
49. Tommaso Caselli, Valerio Basile, Jelena Mitrović, Inga Kartoziya, and Michael Granitzer. I feel offended, don't be abusive! implicit/explicit messages in offensive and abusive language. In *Proceedings of the Twelfth Language Resources and Evaluation Conference*, pages 6193–6202, 2020.
50. Valerio Basile, Cristina Bosco, Elisabetta Fersini, Debora Nozza, Viviana Patti, Francisco Manuel Rangel Pardo, Paolo Rosso, and Manuela Sanguinetti. Semeval-2019 task 5: Multilingual detection of hate speech against immigrants and women in twitter. In *Proceedings of the 13th international workshop on semantic evaluation*, pages 54–63, 2019a.
51. Marcos Zampieri, Shervin Malmasi, Preslav Nakov, Sara Rosenthal, Noura Farra, and Ritesh Kumar. Semeval-2019 task 6: Identifying and categorizing offensive language in social media (offenseval). *arXiv preprint* arXiv:1903.08983, 2019.
52. Antigoni Founta, Constantinos Djouvas, Despoina Chatzakou, Ilias Leontiadis, Jeremy Blackburn, Gianluca Stringhini, Athena Vakali, Michael Sirivianos, and Nicolas Kourtellis. Large scale crowdsourcing and characterization of twitter abusive behavior. In *Proceedings of the international AAAI conference on web and social media*, volume 12, 2018.
53. Victor Sanh, Lysandre Debut, Julien Chaumond, and Thomas Wolf. Distilbert, a distilled version of bert: smaller, faster, cheaper and lighter. *arXiv preprint* arXiv:1910.01108, 2019.
54. Nils Reimers and Iryna Gurevych. Sentence-bert: Sentence embeddings using siamese bert-networks. *arXiv preprint* arXiv:1908.10084, 2019.
55. Mithun Das, Somnath Banerjee, and Animesh Mukherjee. Data bootstrapping approaches to improve low resource abusive language detection for indic languages. In *Proceedings of the 33rd ACM Conference on Hypertext and Social Media*, pages 32–42, 2022.

56. Vikram Gupta, Sumegh Roychowdhury, Mithun Das, Somnath Banerjee, Punyajoy Saha, Binny Mathew, Animesh Mukherjee, et al. Multilingual abusive comment detection at scale for indic languages. *Advances in Neural Information Processing Systems*, 35:26176–26191, 2022.
57. Kaiming He, Xiangyu Zhang, Shaoqing Ren, and Jian Sun. Deep residual learning for image recognition. In *Proceedings of the IEEE conference on computer vision and pattern recognition*, pages 770–778, 2016.
58. Karen Simonyan and Andrew Zisserman. Very deep convolutional networks for large-scale image recognition. *arXiv preprint* arXiv:1409.1556, 2014.
59. Gao Huang, Zhuang Liu, Laurens Van Der Maaten, and Kilian Q Weinberger. Densely connected convolutional networks. In *Proceedings of the IEEE conference on computer vision and pattern recognition*, pages 4700–4708, 2017.
60. Alexey Dosovitskiy, Lucas Beyer, Alexander Kolesnikov, Dirk Weissenborn, Xiaohua Zhai, Thomas Unterthiner, Mostafa Dehghani, Matthias Minderer, Georg Heigold, Sylvain Gelly, Jakob Uszkoreit, and Neil Houlsby. An image is worth 16x16 words: Transformers for image recognition at scale. *ICLR*, 2021.
61. Meng-Hao Guo, Cheng-Ze Lu, Zheng-Ning Liu, Ming-Ming Cheng, and Shi-Min Hu. Visual attention network. *arXiv preprint* arXiv:2202.09741, 2022.
62. Dong Yu and Lin Deng. *Automatic speech recognition*, volume 1. Springer, 2016.
63. Zrar Kh Abdul and Abdulbasit K Al-Talabani. Mel frequency cepstral coefficient and its applications: A review. *IEEE Access*, 2022.
64. Alan V Oppenheim. *Discrete-time signal processing*. Pearson Education India, 1999.
65. Richard A Altes. Detection, estimation, and classification with spectrograms. *The Journal of the Acoustical Society of America*, 67(4):1232–1246, 1980.
66. Chengyi Wang, Yu Wu, Yao Qian, Kenichi Kumatani, Shujie Liu, Furu Wei, Michael Zeng, and Xuedong Huang. Unispeech: Unified speech representation learning with labeled and unlabeled data. In *International Conference on Machine Learning*, pages 10937–10947. PMLR, 2021.
67. Alexei Baevski, Yuhao Zhou, Abdelrahman Mohamed, and Michael Auli. wav2vec 2.0: A framework for self-supervised learning of speech representations. *Advances in neural information processing systems*, 33:12449–12460, 2020.
68. Sanyuan Chen, Chengyi Wang, Zhengyang Chen, Yu Wu, Shujie Liu, Zhuo Chen, Jinyu Li, Naoyuki Kanda, Takuya Yoshioka, Xiong Xiao, et al. Wavlm: Large-scale self-supervised pretraining for full stack speech processing. *IEEE Journal of Selected Topics in Signal Processing*, 16(6):1505–1518, 2022.
69. Shivam Sharma, Firoj Alam, Md Shad Akhtar, Dimitar Dimitrov, Giovanni Da San Martino, Hamed Firooz, Alon Halevy, Fabrizio Silvestri, Preslav Nakov, and Tanmoy Chakraborty. Detecting and understanding harmful memes: A survey. *arXiv preprint* arXiv:2205.04274, 2022.
70. Mithun Das and Animesh Mukherjee. Transfer learning for multilingual abusive meme detection. In *Proceedings of the 15th ACM Web Science Conference 2023*, pages 245–250, 2023b.
71. Jiasen Lu, Dhruv Batra, Devi Parikh, and Stefan Lee. Vilbert: Pretraining task-agnostic visiolinguistic representations for vision-and-language tasks. *Advances in neural information processing systems*, 32, 2019.
72. Liunian Harold Li, Mark Yatskar, Da Yin, Cho-Jui Hsieh, and Kai-Wei Chang. Visualbert: A simple and performant baseline for vision and language. *arXiv preprint* arXiv:1908.03557, 2019.
73. Yen-Chun Chen, Linjie Li, Licheng Yu, Ahmed El Kholy, Faisal Ahmed, Zhe Gan, Yu Cheng, and Jingjing Liu. Uniter: Universal image-text representation learning. In *European conference on computer vision*, pages 104–120. Springer, 2020.
74. Alec Radford, Jong Wook Kim, Chris Hallacy, Aditya Ramesh, Gabriel Goh, Sandhini Agarwal, Girish Sastry, Amanda Askell, Pamela Mishkin, Jack Clark, et al. Learning transferable visual models from natural language supervision. In *International conference on machine learning*, pages 8748–8763. PMLR, 2021.

75. Hao Tan and Mohit Bansal. Lxmert: Learning cross-modality encoder representations from transformers. In *Proceedings of the 2019 Conference on Empirical Methods in Natural Language Processing and the 9th International Joint Conference on Natural Language Processing (EMNLP-IJCNLP)*, pages 5100–5111, 2019.
76. Mithun Das, Rohit Raj, Punyajoy Saha, Binny Mathew, Manish Gupta, and Animesh Mukherjee. Hatemm: A multi-modal dataset for hate video classification. *arXiv preprint* arXiv:2305.03915, 2023.
77. Lingyao Li, Lizhou Fan, Shubham Atreja, and Libby Hemphill. " hot" chatgpt: The promise of chatgpt in detecting and discriminating hateful, offensive, and toxic comments on social media. *arXiv preprint* arXiv:2304.10619, 2023.
78. OpenAI. Introducing chatgpt. https://openai.com/blog/chatgpt, 2023b. Accessed: 2023-04-05.
79. Google. Bard. https://bard.google.com/, 2023. Accessed: 2023-04-05.
80. OpenAI. Gpt-4 technical report, 2023a.
81. Hyung Won Chung, Le Hou, Shayne Longpre, Barret Zoph, Yi Tay, William Fedus, Yunxuan Li, Xuezhi Wang, Mostafa Dehghani, Siddhartha Brahma, et al. Scaling instruction-finetuned language models. *arXiv preprint* arXiv:2210.11416, 2022.
82. Hugo Touvron, Thibaut Lavril, Gautier Izacard, Xavier Martinet, Marie-Anne Lachaux, Timothée Lacroix, Baptiste Rozière, Naman Goyal, Eric Hambro, Faisal Azhar, et al. Llama: Open and efficient foundation language models. *arXiv preprint* arXiv:2302.13971, 2023.
83. Anas Awadalla, Irena Gao, Josh Gardner, Jack Hessel, Yusuf Hanafy, Wanrong Zhu, Kalyani Marathe, Yonatan Bitton, Samir Gadre, Shiori Sagawa, et al. Openflamingo: An open-source framework for training large autoregressive vision-language models. *arXiv preprint* arXiv:2308.01390, 2023.
84. Wenliang Dai, Junnan Li, Dongxu Li, Anthony Meng Huat Tiong, Junqi Zhao, Weisheng Wang, Boyang Li, Pascale Fung, and Steven Hoi. Instructblip: Towards general-purpose vision-language models with instruction tuning, 2023.

# Challenges in Hate Speech Identification

The previous chapter explored the evaluation of hate speech detection models, primarily focusing on their performance on held-out (test) data. While higher metric values suggest more desirable performance, it is crucial to recognize that evaluated metrics alone do not guarantee a robust model. Suppose systematic gaps and biases exist in the training data. In that case, models may superficially excel on corresponding test sets by learning data artifacts rather than understanding the actual task they were trained for. Further, understanding why machine learning models classify specific content as hate speech is essential. This comprehension helps us identify when a model fails and guides us to implement a robust model. The chapter is organized into three parts.

- **Pitfalls of model evaluation**: We illustrate how incorrect data distribution can lead to inaccurate observations.
- **Bias and fairness**: We explore the concepts of bias and fairness and their impact on model performance.
- **Explainability**: We investigate explainability and how interpretable models can contribute to better judgments in determining the hateful nature of a post.

## 4.1 Pitfalls of Model Evaluation

### 4.1.1 Caveats of High Performance Values

In the pursuit of enhancing the performance of hate speech detection models, researchers and practitioners often face the challenge of feature extraction. This critical step involves transforming raw data into a format suitable for machine learning algorithms. While achieving high accuracy or F1 scores for a model may initially seem promising, it is crucial to scrutinize the process leading up to the model's evaluation. Incorrect feature extractions or oversampling techniques employed before the train-test split can artificially inflate performance metrics. The model may excel during evaluation, displaying impressive accuracy or F1 scores on the test set. However, this performance can be misleading, as it might result from the model learning from overemphasized or incorrect features that do not generalize well to real-world scenarios. The true test of a hate speech detection model lies in its real-world applicability. When deployed in practical settings, relying solely on metrics obtained from incorrectly handled features may lead to poor performance. The model might struggle to discern nuanced instances of hate speech or exhibit unexpected behavior when faced with data outside the oversampled or incorrectly extracted feature distribution. Let us explore two methodological issues [1] that past researchers indulged into while developing hate speech detection models.

### 4.1.1.1 Extracting Features Using the Entire Dataset

Badjatiya et al. [2] conducted a comprehensive investigation into hate speech detection, exploring various traditional machine learning models, deep neural models, and hybrid approaches. Their most promising outcome, achieving an impressive F1 score of 93%, involved a hybrid model combining an RNN for word embeddings and a decision tree for text classification. The methodology comprised two distinct phases:

*Phase 1*: The authors devised an architecture having an embedding Layer, an LSTM network, and a fully connected layer with three neurons plus a softmax activation to predict classes like 'sexist,' 'racist,' and 'non-hate.' This architecture was trained end-to-end on labeled tweets, utilizing cross-entropy as the loss function and the Adam optimizer with default learning rates. Although capable of directly predicting the class of an input text, this architecture primarily served as a feature extractor for Phase 2.

*Phase 2*: Leveraging the embeddings learned in Phase 1, the authors processed input tweets as sequences of words, producing a sequence of vectors for each word. These vectors were then averaged to create a single vector, which was fed into a gradient-boosted decision tree (GBDT) for classification. This yielded micro-average and macro-average F1 scores of 94% and 93%, respectively (see Table 4.1).

However, the methodological concern in this case is the features obtained in Phase 1 by considering the complete labeled dataset. Specifically, let $T$ be the labeled dataset and assume it is divided as $T = T_{\text{train}} \cup T_{\text{test}}$. The entire labeled dataset ($T = T_{\text{train}} \cup T_{\text{test}}$) is used for the feature extraction. For the classification phase, the set $T_{\text{train}}$ is used to train the classifier, and $T_{\text{test}}$ is used to evaluate it; when one sees the complete pipeline, the set $T_{\text{test}}$ is simultaneously used to train and validate the entire architecture. This may lead to an artificial increase in the performance of the model. Upon re-evaluating [1] the method by extracting features solely from $T_{\text{train}}$ and training the GBDT classifier on the same set, the reported

4.1 Pitfalls of Model Evaluation

**Table 4.1** Replication of the state-of-the-art results [2, 3]

| Method | Class | Prec. | Rec. | F1 |
|---|---|---|---|---|
| Badjatiya et al. [2] Emb. over all dataset | Neither | 95.5 | 96.8 | 96.1 |
| | Racist | 94.5 | 93.5 | 94.0 |
| | Sexist | 91.2 | 87.5 | 89.3 |
| | Micro avg. | 94.6 | 94.6 | 94.6 |
| | Macro avg. | 93.7 | 92.6 | 93.1 |
| Agrawal and Awekar [3] Oversamp. all dataset | Neither | 95.1 | 91.7 | 93.4 |
| | Racist | 94.9 | 96.0 | 95.4 |
| | Sexist | 92.5 | 97.0 | 94.6 |
| | Micro avg. | 94.4 | 94.4 | 94.4 |
| | Macro avg. | 94.2 | 94.9 | 94.5 |

**Table 4.2** Replication of the state-of-the-art results taking into account the methodological issues [1]

| Method | Class | Prec. | Rec. | F1 |
|---|---|---|---|---|
| Badjatiya et al. [2] Emb. over train set | Neither | 82.3 | 94.7 | 88.1 |
| | Racist | 78.0 | 64.0 | 70.2 |
| | Sexist | 84.5 | 47.8 | 60.9 |
| | Micro avg. | 82.3 | 82.1 | 80.7 |
| | Macro avg. | 81.6 | 68.9 | 73.1 |
| Agrawal and Awekar [3] Oversamp. train set | Neither | 90.3 | 86.5 | 88.3 |
| | Racist | 69.6 | 81.3 | 75.0 |
| | Sexist | 74.0 | 77.4 | 75.5 |
| | Micro avg. | 84.7 | 84.1 | 84.3 |
| | Macro avg. | 78.0 | 81.7 | 79.6 |

metrics experienced a significant decrease. F1 scores for each class dropped from 96.1 to 88.1 (neither), 94.0 to 70.2 (racist), and 89.3 to 60.9 (sexist), leading to a macro-average F1 drop of 20 points (see Table 4.2), i.e., from 93.1 to 73.1.

#### 4.1.1.2 Oversampling Before the Train-Test Split

Agrawal and Awekar [3] explored various models akin to Badjatiya et al. [2]. They found that deep neural models consistently outperformed other methods. They presented results for

CNNs, LSTMs, BiLSTMs, and BiLSTMs with attention. While the results for all architectures were similar, [1] focused on reproducing only the BiLSTM results to avoid reporting unnecessary results. The architecture closely mirrors *Phase 1* in Badjatiya et al. [2], featuring layers of embedding → recurrent layers → fully connected layers → softmax, yielding a macro F1 score of 94.5.

A critical insight from Agrawal and Awekar [3] was the impact of oversampling. Notably, they addressed the class imbalance problem by oversampling the class with fewer examples. However, a critical issue arose as they applied oversampling to the entire dataset before the train-test split. Oversampling should ideally be limited to the training data to prevent data leakage, where some data points end up in both training and test sets during random splitting. When Arango et al. [1] replicated their approach, incorporating oversampling after the train-test split, F1 scores for each class decreased from 93.4 to 88.3 (neither), 95.4 to 75.0 (racist), and 94.6 to 75.5 (sexist), which resulted in a notable macro-average F1 drop of 15 points (see Tables 4.1 and 4.2), from 94.5 to 79.6.

### 4.1.2 Impact of User Distribution on Model Generalization

While addressing feature extraction techniques and oversampling, we have observed a decline in the model's performance. Simultaneously, the distribution of users also emerges as a critical factor influencing the classification process. Most hate speech datasets are constructed by crawling and labeling posts from various social media platforms. However, if a significant portion of the collected hateful posts originates from a specific subset of users within the platform then this might lead to misleading observations. In such a case, the classification model might only pick up the traits associated with these users instead of developing a nuanced understanding of when a post should genuinely be classified as hateful.

The Waseem and Hovy's dataset [4], for instance, displays a significant imbalance, with only 20% of tweets labeled as 'sexist' and 12% labeled as 'racist'. Moreover, a single user generates 44% of 'sexist' comments, and a single user makes 96% of 'racist' comments [1]. The impact becomes evident when splitting the dataset into training and test sets with a strict, non-overlapping user policy. Arango et al. [1] conducted an experiment to ensure no overlapping users between the train and test sets. The motivation behind this task is to prevent the classifier from learning any secondary task (e.g., identifying users) instead of the primary task it is trained for (e.g., identifying hate speech). The results are reported in Table 4.3. Considering the inherent imbalance in the original dataset, deliberately preventing the classifier from learning user-specific patterns by ensuring distinct users in both training and testing sets results in a dramatic drop in model performance. Hence the Waseem and Hovy dataset [4], which has been widely used, may not be optimal for solving the hate speech detection problem. Therefore, when crafting datasets for hate speech detection, it is crucial to ensure that datasets do not disproportionately include posts from particular users.

**Table 4.3** Results obtained by partitioning the Waseem and Hovy dataset [4] into train and test sets considering the user distribution (no overlapping users between train and test sets) [1]

| Method | Class | Prec. | Rec. | F1 |
|---|---|---|---|---|
| Badjatiya et al. [2] | None | 49.6 | 93.4 | 64.3 |
| | Hateful | 68.8 | 15.4 | 23.5 |
| | Micro avg. | 63.8 | 54.1 | 46.1 |
| | Macro avg. | 59.2 | 54.4 | 43.9 |
| Agrawal and Awekar [3] | None | 47.5 | 98.0 | 63.0 |
| | Hateful | 75.3 | 03.5 | 06.7 |
| | Micro avg. | 62.3 | 48.4 | 35.1 |
| | Macro avg. | 61.4 | 50.8 | 34.9 |

## 4.2 Bias and Fairness

While we have highlighted how incorrect data distribution can result in inaccurate observations, additional challenges persist within the hate speech detection system that warrant consideration, especially concerning bias and fairness.

### 4.2.1 Bias

Bias in hate speech detection refers to prejudice or unfairness in the model's decision-making process. Such bias can lead to incorrect or discriminatory labeling of content as hate/non-hate speech, disproportionately impacting certain marginalised groups such as African Americans, Jews, Muslims, women, and more [5, 6]. Bias can lead to significant consequences, such as silencing legitimate voices, perpetuating harmful stereotypes, and reinforcing existing inequalities. Bias in AI systems can result from various sources, including data, algorithms, and the design and implementation of the system. Here are several types of biases commonly observed in hate speech detection systems.

- **Data bias** [6, 7]

    - *Underrepresentation*: When certain groups or perspectives are insufficiently represented in the training data, the model may struggle to accurately detect hate speech targeting those groups.
    - *Overrepresentation*: Conversely, if certain groups are overrepresented, the model might be more proficient at identifying hate speech against those groups while potentially neglecting other forms of hate speech.

- **Annotator bias** [6, 8]

  - *Subjectivity in labeling*: Human annotators may have subjective interpretations of hate speech, leading to inconsistencies in labeling.
  - *Cultural sensitivity*: Annotators may not be familiar with the cultural context or language nuances of certain groups, resulting in misinterpretations.

- **Algorithmic bias** [9]

  - *Biases in word vectors* [10]: Word embeddings trained on biased datasets can carry and perpetuate biases. Certain words may be associated with specific groups or sentiments, affecting the model's predictions.
  - *Contextual bias* [5]: Models may struggle to comprehend the context in which certain words or phrases are used, leading to false positives or negatives.

- **Socio-demographic bias** [11]

  - *Demographic disparities* [12, 13]: The model may exhibit biases related to gender, race, ethnicity, or other socio-demographic factors present in the training data.
  - *Language and dialectal bias* [14]: Models may perform better on certain dialects or languages, leading to disparities in detection accuracy.

- **Temporal bias** [15]

  - *Outdated training data* [16]: If the training data does not reflect current language trends or societal changes, the model may struggle to adapt to evolving forms of hate speech.
  - *Changing contexts* [17]: The evolving nature of language and cultural contexts can lead to biases if the model is not regularly updated.

Understanding and addressing these biases is crucial for developing more ethical and effective hate speech detection systems; otherwise, the model's predictions may disproportionately harm certain groups, exacerbating existing societal inequalities.

### 4.2.2 Fairness

Fairness in the context of artificial intelligence and machine learning refers to ensuring that algorithmic decision-making systems do not systematically and unfairly discriminate against or favor specific groups or individuals [18]. Fairness is closely related to addressing bias in AI systems but goes beyond recognizing and mitigating bias. It ensures that AI systems treat all individuals and groups equitably and impartially, regardless of their personal characteristics or social status.

Let us understand fairness in more detail using an example related to hate speech detection. Suppose we have an AI system designed to detect hate speech in user-generated content on a social media platform. To ensure fairness, we need to consider various aspects as follows.

- *Data fairness* [19]: Our training dataset should be diverse and representative of different demographics, cultures, and languages. It should not favor one group over another, and we should address the underrepresentation of minority groups.
- *Algorithmic fairness* [20]: The AI model should be designed and trained to avoid favoring any particular group. This means the model should not disproportionately flag content from certain groups as hate speech while ignoring similar content from others.
- *Evaluation fairness* [21]: When assessing the model's performance, use fairness-aware evaluation metrics that consider false positives and false negatives across different groups. Ensure that the system does not optimize for one group at the expense of others.
- *Transparency* [22]: Ensure that the decision-making process of the AI system is transparent and explainable so that users and stakeholders can understand why a particular decision was made.
- **Temporal fairness** [23]: Temporal fairness examines the system's consistency and performance over time. It requires ensuring that the system's accuracy and bias do not fluctuate disproportionately over different time periods, adapting to changing language dynamics and societal contexts.

Fairness in hate speech detection is not just an ethical imperative but also a practical necessity for effective and trustworthy AI systems. By prioritizing fairness in the design, development, and deployment of these technologies, we can create AI systems that promote equality, protect vulnerable groups, and contribute to a more just and equitable society.

### 4.2.3 Fairness Versus Bias

While bias and fairness [24, 25] are closely related concepts, they are distinct. Bias refers to the tendency or inclination to favor one perspective or set of ideas over others. On the other hand, fairness encompasses the broader notion of ensuring that individuals and groups are treated equitably.

In the context of AI, bias can lead to unfair outcomes, but it is not necessarily indicative of intentional discrimination. For instance, a hate speech detection system biased towards a particular dialect may not intentionally target speakers of that dialect. However, the bias can still result in unfair treatment. Fairness, on the other hand, goes beyond simply avoiding bias. It requires actively designing and implementing AI systems that treat all individuals and groups equitably. This may involve using diverse training data, employing fairness metrics, and overcome human oversight to ensure fair outcomes.

### 4.2.4 Measuring Bias in Hate Speech Detection Models

Borkan et al. [26] introduces a comprehensive suite of threshold-agnostic metrics designed to offer a nuanced understanding of unintended bias. These metrics consider the diverse ways in which a classifier's score distribution may vary across specified groups. Conventional metrics like accuracy fall short in measuring bias as they neglect the distribution of scores across different demographic groups. In response to this challenge, the authors developed a set of metrics, including subgroup AUC, background positive subgroup negative (BPSN) AUC, background negative subgroup positive (BNSP) AUC, and generalized mean of bias AUCs.

- *Subgroup AUC*: Here, we select the toxic and normal posts from the test set which mention the community under consideration. The ROC-AUC score of this set will provide us with the subgroup AUC for a community. This metric measures the model's ability to separate the toxic and normal comments in the context of the community (e.g., Asians, homosexuals, etc.). A higher value means that the model is doing a good job at distinguishing the toxic and normal posts specific to the community.
- *BPSN AUC*: Here, we select the normal posts which mention the community and the toxic posts which do not mention the community, from the test set. The ROC-AUC score of this set will provide us with the BPSN AUC for a community. This metric measures the false positive rates of the model with respect to a community. A higher value means that a model is less likely to confuse between the normal post that mentions the community with a toxic post which does not.
- *BNSP AUC*: Here, we select the toxic posts which mention the community and the normal posts which do not mention the community, from the test set. The ROC-AUC score for this set will provide us with the BNSP AUC for a community. The metric measures the false negative rates of the model with respect to a community. A higher value means that the model is less likely to confuse between a toxic post that mentions the community with a normal post without one.
- *GMB AUC*: This metric was introduced by the Google Conversation AI Team as part of their Kaggle competition.[1] This metric combines the per-identity bias AUCs into one overall measure as $M_p(m_s) = \left(\frac{1}{N}\sum_{s=1}^{N} m_s^p\right)^{\frac{1}{p}}$ where, $M_p$ = the $p$th power-mean

---

[1] https://www.kaggle.com/c/jigsaw-unintended-bias-in-toxicity-classification/overview/evaluation.

function, $m_s$ = the bias metric $m$ calculated for subgroup $s$ and $N$ = number of identity subgroups.

### 4.2.5 How to Make Hate Speech Detection Models More Fair and Less Biased

Developing fair and unbiased hate speech detection models is crucial to ensure that these systems are not perpetuating harmful biases and unfairly targeting marginalized groups. Here are some strategies [27, 28] to address bias in hate speech detection models.

- **Data collection and curation**

  - *Diverse and representative training data*: Ensure that the training data used to train hate speech detection models is diverse and representative of various demographic groups and language variations. This helps the model learn from a broad range of examples and reduces the risk of biases.
  - **Careful data annotation**: Employ annotators with diverse backgrounds sensitive to different social and cultural contexts to mitigate subjectivity in hate speech labeling. Further, engage experts from different backgrounds to conduct a thorough manual review of the data.
  - *Data augmentation*: Synthesize additional training data to address imbalances and under-representation, especially for minority groups.

- **Algorithmic design and training**

  - *Context-aware models*: Develop models that consider the broader context of the text, including surrounding words, sentiment, and speaker identity, to avoid misinterpretations.
  - *Fairness constraints*: During model training, enforce fairness constraints like equality of opportunity or equality of odds to ensure equitable outcomes across different groups.
  - *Examine feature importance*: Analyze which features the model relies on to make predictions and adjust to prioritize relevant signals over potentially biased ones.

- **Human oversight and evaluation**

  - **Regular human review**: Regularly review and evaluate the model's performance on real-world data to identify potential biases and adjust the model accordingly.

conducting a global explainability analysis; it might show a strong reliance on specific types of offensive language, mentions of particular identity groups, or the frequency of specific keywords.

For instance, a hate speech detection model might consistently classify text instances that contain the word 'hate' as hate speech. Global explainability techniques, such as feature importance analysis, can reveal that this word is highly important in the model's predictions. This information suggests that the model is heavily relying on the presence of specific words to determine hate speech, which could potentially lead to false positives or false negatives. By identifying the most influential features and understanding their impact on the model's predictions, developers can refine the model to improve its accuracy and reduce bias.

Global explainability enables a platform to understand the prevailing themes and challenges related to hate speech on its network. It helps developers and moderators identify systemic issues, such as potential biases or areas where the model may struggle, allowing for informed adjustments and improvements. Moreover, global explainability is instrumental in promoting transparency and accountability on a broader scale, contributing to the responsible deployment of hate speech detection models across large and diverse user bases.

Table 4.4 summarizes the key differences between global and local explainability.

In practice, both global and local explainability are essential for understanding and interpreting hate speech detection models. Global explainability provides a high-level overview of the model's behavior, while local explainability provides detailed insights into specific predictions. By combining these two approaches, we can understand how the model works and why it makes the predictions it does.

### 4.3.1.2 Self-explaining Versus Post-hoc Approaches

Whether the explanation is local or global, variations exist in how explanations emerge within the prediction context. The difference lies in whether these explanations unfold seamlessly during the prediction process or necessitate post-processing once the model has made its prediction. A *self-explaining* methodology, which may also be referred to as directly interpretable [31], generates the explanation concurrently with the prediction. It draws on information emitted by the model during the prediction process. Decision trees and rule-based models are examples of global self-explaining models, while feature saliency techniques such as attention maps are illustrations of local self-explaining models. In contrast, a *post-hoc* approach mandates an additional operation after predictions are made. LIME [32] serves as an illustrative example, generating a local explanation using a surrogate model applied subsequent to the prediction pipeline [29].

## 4.3 Explainability

**Table 4.4** Differences between global and local explainability

| Aspect | Global explainability | Local explainability |
|---|---|---|
| Focus | Understands overall behavior and patterns across the dataset | Examines the rationale behind individual predictions |
| Scope | Provides a broad, holistic view | Offers insights into a specific instance or data point |
| Purpose | Identifies general trends, biases, and model tendencies | Reveals why a particular prediction was made for an input |
| Application | Useful for refining models, addressing biases at a systemic level | Valuable for improving accuracy on a case-by-case basis |
| Examples | Identifying common features influencing predictions | Explaining why a specific post was classified as hate speech |
| Scale | Analyzes the entire dataset or a significant subset | Focuses on an individual instance or a small set of instances |
| Decision-making Impact | Influences high-level model adjustments and improvements | Informs fine-tuning for specific instances or user feedback |
| Application | Model development and refinement | User dispute resolution and moderation decisions |
| Limitations | May not reveal the exact reasons for individual predictions | May require more computational resources |
| Relevance to hate speech detection | Can help identify general trends and biases in hate speech detection | Can provide detailed explanations for why specific text instances are classified as hate speech |

### 4.3.2 Common Techniques for Explainability

While several explainability techniques have been disseminated within the NLP research community, we delve into a selection of prevalent methods widely adopted by NLP practitioners.

#### 4.3.2.1 Feature Importance

Feature importance assesses the relative contribution of each feature to the model's prediction. It reveals which features have the most influence on the model's decision to classify a text as hate speech or not. These approaches can be constructed based on a variety of features [32], such as manual features derived from feature engineering [33], lexical features, including words/tokens and n-grams [34], or latent features [35] acquired through neural networks. In the context of hate speech detection, feature importance can help in—

- **Identifying biases**: Feature importance helps to determine whether the model relies on biased features, such as certain keywords or demographic information, for its analysis. This helps to mitigate potential discrimination and promote fairness.
- **Improving model performance**: By understanding the key features contributing to hate speech detection, developers can focus on more relevant features during training and improve model accuracy.
- **Understanding model behavior**: Feature importance provides insights into what the model considers offensive and helps explain its decision-making process.

#### 4.3.2.2 LIME (Local Interpretable Model-Agnostic Explanations)

LIME [32] is a technique used to provide local explanations for individual predictions made by any machine learning classifier. The key idea is to locally approximate the black-box model's decision boundary by creating a simple, interpretable surrogate model [29]. LIME works by perturbing the input features of a specific instance, observing the model's predictions on these perturbed instances, and then fitting a locally interpretable model to approximate the complex model's behavior. LIME provides local explanations for individual predictions, allowing for a deeper understanding of why a specific prediction was made.

**LIME Algorithm**

- **Select an instance**: Choose a specific instance for which we want to generate an explanation. For our case, let us consider: we want to understand why the following text—*All members of Group X are criminals and should be deported immediately!*—has been classified as hate speech.
- **Perturb the instance**: Create perturbed versions of the chosen instance by slightly modifying the text *Group X members are criminals and should be deported immediately!, All individuals in Group X are criminals and should be deported immediately!* etc.
- **Model prediction**: Obtain hate speech predictions from the original model for both the original and perturbed instances.
- **Build a surrogate model**: Fit a simple, interpretable model (e.g., linear regression, decision tree) to the perturbed instances, using their modified text and the corresponding hate speech predictions from the original model.
- **Interpret the surrogate model**: Analyze the decision rules of the surrogate model to understand which words or phrases contributed the most to the hate speech classification. The surrogate model provides a local approximation of the original model's decision boundary for this specific post.

## 4.3 Explainability

**Example interpretation**: The surrogate model might identify that the presence of phrases like—*Group X members*, *criminals*, and *deported immediately* strongly correlates with the hate labels. This may reveal that the model flagged the text as hate speech due to the association of a specific group with criminality and the call for immediate deportation.

Overall, LIME is a powerful **model-agnostic** tool for explaining the predictions of deep neural models. It is an important tool for debugging and improving models, providing feedback to users, and promoting transparency and trust.

### 4.3.2.3 SHAP (SHapley Additive exPlanations)

SHAP [36] is another **model-agnostic** technique for explaining the output of machine learning models by assigning a value (Shapley value [37]) to each feature, indicating its contribution to a particular prediction. Each Shapley value represents the impact of the feature in generating the prediction delivered by the model. Shapley values are a concept from game theory, developed initially as a measure to distribute a reward fairly among a set of players contributing to a particular outcome. In the context of machine learning models, the players involved are the input features, and the outcome is the model's decision; Shapley values attribute an importance score to each part of the input [38].

**SHAP algorithm**

– **Baseline prediction**: Assume the baseline prediction is the average prediction of the model on the training dataset.
– **Create feature coalitions**: Consider all possible subsets of words or phrases in the hate speech text.
– **Calculate marginal contributions**: For each word or phrase, calculate its marginal contribution to every coalition it belongs to. The marginal contribution is the difference between the model's prediction with the word or phrase and the model's prediction without it for a specific coalition.
– **Calculate Shapley values**: For each word or phrase, calculate its Shapley value by averaging its marginal contributions over all possible coalitions. This ensures a fair distribution of the overall impact among the words or phrases.
– **Assign feature importance**: The obtained Shapley values represent the importance of each word or phrase in the hate speech text for the specific instance. Positive values indicate elements that contribute to the hate speech classification, while negative values suggest elements that may mitigate the likelihood of hate speech.
– **Summation and consistency**: Verify that the sum of Shapley values for all words or phrases equals the difference between the model's prediction for the hate speech text and the baseline prediction. This ensures the additivity property and consistency in the explanation.

- **Interpretability**: Use the Shapley values to understand which words or phrases contributed the most to the model's prediction that the text is hate speech. Positive values may indicate terms strongly associated with hate speech, offering insights into the decision-making process.

### 4.3.3 Evaluating Rationales

Now that we have a fair understanding of how machine learning models interpret posts to render judgments on instances fed into the model, it becomes crucial to establish metrics that measure the alignment between the model's judgment (rationale) and human judgment. Consider the following hate speech text as an example: 'Muslims are the worst of all religious communities'. The annotator identifies the tokens '*Muslims*', '*worst*', and '*religious*' as rationales to label the post as hateful. Suppose the machine learning model also interprets the post as hateful based on these tokens. In that case, we can assert that the machine learning model interprets why a post is hateful similar to a human.

DeYoung et al. [39] proposed metrics that aim to capture how well the rationales provided by models align with human rationales and also how faithful these rationales are (i.e., the degree to which provided rationales influenced the corresponding predictions) and these metrics are named as **plausibility** and **faithfulness**.

#### 4.3.3.1 Plausibility
Plausibility can be approximated using metrics that measure the alignment of an explanation with human intuition and common sense. In other words, a plausible explanation for a natural language classification should be understandable and convincing to humans, even if they are not experts in the underlying AI model or natural language processing (NLP) domain. To assess plausibility, [39] employ metrics for both discrete and soft selection. In the discrete case, intersection-over-union (IOU) is computed on a token level, defined as the size of the overlap of tokens covered by two spans divided by the size of their union. A prediction is deemed a match if the overlap with any ground truth rationale exceeds 0.5. These partial matches contribute to calculating an F1-score (IOU F1). In addition, [39] measures token-level precision and recall, leveraging these to compute token-level F1 scores (token F1). For continuous or soft token scoring models, metrics consider token rankings, rewarding models for assigning higher scores to marked tokens. In particular, [39] utilize the area under the precision-recall curve (AUPRC) constructed by sweeping a threshold over token scores to evaluate the performance of such models.

## 4.3 Explainability

### 4.3.3.2 Faithfulness

Faithfulness can be approximated using metrics that assess the accuracy and relevance of an explanation in reflecting the true reasons for a model's prediction. One approach is to compare the explanation to a ground truth representation of the model's decision-making process, which may be derived from the model's internal structure or from expert knowledge. To measure faithfulness, [39] report two metrics: *comprehensiveness* and *sufficiency*.

- **Comprehensiveness**: To measure comprehensiveness, [39] create a contrast example $\tilde{x}_i$, for each post $x_i$, where $\tilde{x}_i$ is calculated by removing the predicted rationales $r_i$ from $x_i$. Let $m(x_i)_j$ be the original prediction probability provided by a model $m$ for the predicted class $j$. Then DeYoung et al. [39] define $m(x_i \backslash r_i)_j$ as the predicted probability of $\tilde{x}_i$ ($= x_i \backslash r_i$) by the model $m$ for the class $j$. The expectation is, the model prediction to be lower on removing the rationales. This can measured as follows—comprehensiveness $= m(x_i)_j - m(x_i \backslash r_i)_j$. A high value of comprehensiveness implies that the rationales were influential in the prediction.
- **Sufficiency** measures the degree to which extracted rationales are adequate for a model to make a prediction. Thus this can measured as follows—sufficiency $= m(x_i)_j - m(r_i)_j$.

### 4.3.4 Explainable Hate Speech Detection

In the preceding section, we explored the concept of explainability and delved into various techniques aimed at elucidating the decision-making processes of models. Now, let us attempt to have a deeper understanding of how existing hate speech detection models truly operate in formulating their predictions. It is observed that, at times, although models yield accurate predictions, the explanations offered by them often diverge from human-made interpretations. Further, even when trained on the same datasets, different models derive distinct 'rationales' for their predictions. This discrepancy may arise from models learning varied features or patterns from the training data. Extensive research has been conducted to enhance the understanding of hate speech detection models and subsequently refine the quality of explanations they provide.

**HateXplain** [30] serving as the pioneering benchmark dataset for explainable hate speech detection, comprises an extensive collection of text posts annotated with three key aspects: fundamental classification (categorized as hate, offensive, or normal), target community identification, and rationales—signifying the specific segments of the post upon which the labeling decision is grounded. The underlying expectation is that incorporating human judgment into the model during training should foster enhanced learning, thereby resulting in improved performance and a reduction of unintended bias directed toward specific target communities.

**Table 4.5** Examples of rationale-based annotation. The ==highlighted== portion of the text represents the annotator's rationale [30]

| Text | Dad should have told the ==muzrat whore== to ==fuck off==, and went in anyway |
|---|---|
| Label | Hate |
| Targets | Islam |
| Text | A ==nigress too dumb to fuck== has a scant chance of understanding anything beyond the size of a dick |
| Label | Hate |
| Targets | Women, African |
| Text | Twitter is full of tween ==dikes== who think they're superior because of ==''muh oppression.''== News flash: No one gives a shit |
| Label | Offensive |
| Targets | Homosexual |

To annotate rationales, each post labeled as hateful or offensive underwent a secondary annotation process, during which annotators identified and highlighted segments justifying the assigned class. Two to three annotators provided rationale explanations for each post. Notably, the average number of tokens per post in the dataset was found to be 23.42. Out of these, the average number of tokens highlighted per post was 5.48 for offensive speech and 5.47 for hate speech. Table 4.5 shows examples of hate speech text annotations based on rationales.

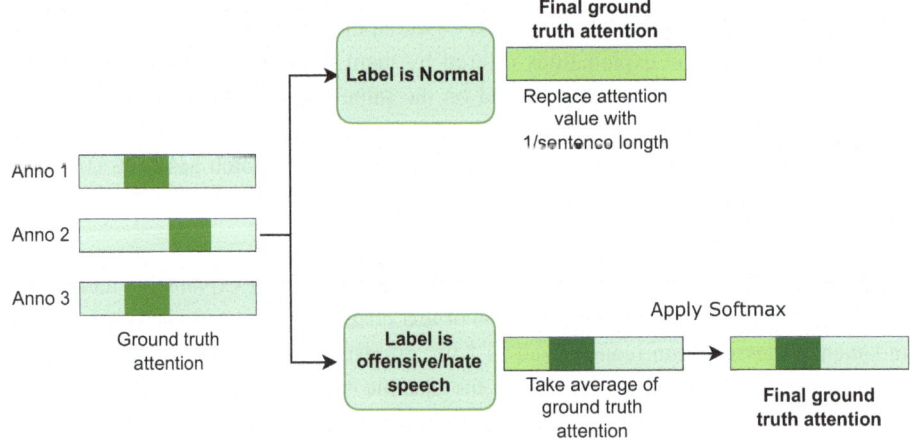

**Fig. 4.1** Ground truth attention [30]

## 4.3 Explainability

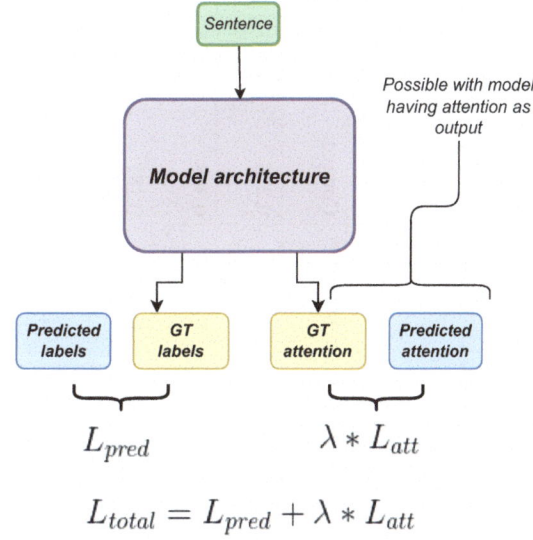

**Fig. 4.2** Representation of the general model architecture showing how the attention of the model is trained using the ground truth (GT) attention. λ controls how much effect the attention loss has on the total loss [30]

$$L_{total} = L_{pred} + \lambda * L_{att}$$

To establish ground truth attention for posts labeled with hate speech or offensive tags, each rationale was transformed into a Boolean attention vector, indicating tokens with a value of 1. The average of these attention vectors generated a common ground truth attention vector for each post. Attention vectors from attention-based models typically sum to 1. To address potential issues, such as low differences between rationale and non-rationale token values, a temperature parameter ($\tau$) in the softmax function has been introduced. This parameter, tuned using the validation set, concentrates the probability distribution on the rationales. In cases where the post is labeled as normal, attention vectors are disregarded, and each element in the ground truth attention is replaced with 1/(sentence length) to represent a uniform distribution. Figure 4.1 illustrates this computation.

To illustrate the benefits of rationale-based annotation, several models such as BiRNN, BERT have been explored with two versions, one where the models are trained using the ground truth class labels only (i.e., hate speech, offensive speech, and normal) and the other, where the models are trained using the ground truth attention and class labels, as shown in Fig. 4.2.

Table 4.6 shows the performance of the explored models. The authors observe that models which utilize the human rationales as part of the training (BiRNN-**HateXplain [LIME & Attn]**, BERT-**HateXplain [LIME & Attn]**)[2]) are able to perform slightly better in terms of the performance metrics. BiRNN-**HateXplain [LIME & Attn]** has improved score for all plausibility metrics and comprehensiveness as compared to BiRNN-**Attn [LIME & Attn]**. In case of BERT-**HateXplain [LIME]**, the faithfulness scores improved as compared to

---

[2] <model>-**HateXplain** denotes the models where supervised attention using ground truth attention vector was used.

**Table 4.6** Model performance results. To select the tokens for explainability calculation, we used attention and LIME methods [30]

| Model [Token method] | Performance | | Bias | | | | Explainability | | | | Faithfulness | |
|---|---|---|---|---|---|---|---|---|---|---|---|---|
| | Acc.↑ | Macro F1↑ | AUROC↑ | GMB-Sub.↑ | GMB-BPSN↑ | GMB-BNSP↑ | Plausibility | | | Comp.↑ | Suff.↓ |
| | | | | | | | IOU F1↑ | Token F1↑ | AUPRC↑ | | |
| BiRNN-Attn [Attn] | 0.621 | 0.614 | 0.795 | 0.653 | 0.662 | 0.668 | 0.167 | 0.369 | 0.643 | 0.278 | 0.001 |
| BiRNN-Attn [LIME] | 0.621 | 0.614 | 0.795 | 0.653 | 0.662 | 0.668 | 0.162 | 0.386 | 0.650 | 0.308 | −0.075 |
| BiRNN-HateXplain [Attn] | 0.629 | 0.629 | 0.805 | 0.691 | 0.636 | 0.674 | **0.222** | **0.506** | **0.841** | 0.281 | 0.039 |
| BiRNN-HateXplain [LIME] | 0.629 | 0.629 | 0.805 | 0.691 | 0.636 | 0.674 | 0.174 | 0.407 | 0.685 | 0.343 | −0.075 |
| BERT [Attn] | 0.690 | 0.674 | 0.843 | 0.762 | 0.709 | 0.757 | 0.130 | 0.497 | 0.778 | 0.447 | 0.057 |
| BERT [LIME] | 0.690 | 0.674 | 0.843 | 0.762 | 0.709 | 0.757 | 0.118 | 0.468 | 0.747 | 0.436 | 0.008 |
| BERT-HateXplain [Attn] | **0.698** | **0.687** | **0.851** | **0.807** | **0.745** | **0.763** | 0.120 | 0.411 | 0.626 | 0.424 | 0.160 |
| BERT-HateXplain [LIME] | **0.698** | **0.687** | **0.851** | **0.807** | **0.745** | **0.763** | 0.112 | 0.452 | 0.722 | **0.500** | 0.004 |

other BERT models. However, the plausibility scores decreased. Similar to performance, models which utilize the human rationales as part of the training are able to perform better in reducing the unintended model bias for all the bias metrics. The authors observe that the presence of community terms within the rationales is effective in reducing the unintended bias. They further note that models such as BERT-**HateXplain [LIME & Attn]**, which have the best scores in terms of performance metrics and bias, do not perform well in terms of plausibility metrics. In fact, BERT-**HateXplain [Attn]** has the worst score for sufficiency as compared to other models. BERT-**HateXplain [LIME]** seems to be the best model for comprehensiveness metric. For plausibility metrics, the authors observe BiRNN-**HateXplain [Attn]** to have the best scores. For sufficiency, CNN-GRU seems to be doing the best [30]. For the token method, LIME seems to be generating more faithful results as compared to attention. These are in agreement with DeYoung et al. [39]. Overall, we observe that a model's performance metric alone is not enough. Models with slightly lower performance but much higher scores for plausibility and faithfulness should be preferred depending on the task at hand.

## References

1. Aymé Arango, Jorge Pérez, and Barbara Poblete. Hate speech detection is not as easy as you may think: A closer look at model validation (extended version). *Information Systems*, 105:101584, 2022.
2. Pinkesh Badjatiya, Shashank Gupta, Manish Gupta, and Vasudeva Varma. Deep learning for hate speech detection in tweets. WWW, pages 759–760, 2017.
3. Sweta Agrawal and Amit Awekar. Deep learning for detecting cyberbullying across multiple social media platforms. In *European conference on information retrieval*, pages 141–153. Springer, 2018.
4. Zeerak Waseem and Dirk Hovy. Hateful symbols or hateful people? predictive features for hate speech detection on twitter. In *Proceedings of the NAACL student research workshop*, pages 88–93, 2016.
5. Maarten Sap, Dallas Card, Saadia Gabriel, Yejin Choi, and Noah A Smith. The risk of racial bias in hate speech detection. In *Proceedings of the 57th annual meeting of the association for computational linguistics*, pages 1668–1678, 2019.
6. Dirk Hovy and Shrimai Prabhumoye. Five sources of bias in natural language processing. *Language and Linguistics Compass*, 15(8):e12432, 2021.
7. Michael Wiegand, Josef Ruppenhofer, and Thomas Kleinbauer. Detection of Abusive Language: the Problem of Biased Datasets. In Jill Burstein, Christy Doran, and Thamar Solorio, editors, *Proceedings of the 2019 Conference of the North American Chapter of the Association for Computational Linguistics: Human Language Technologies, Volume 1 (Long and Short Papers)*, pages 602–608, Minneapolis, Minnesota, June 2019. Association for Computational Linguistics. URL https://aclanthology.org/N19-1060.
8. Hala Al Kuwatly, Maximilian Wich, and Georg Groh. Identifying and measuring annotator bias based on annotators' demographic characteristics. In *Proceedings of the fourth workshop on online abuse and harms*, pages 184–190, 2020.
9. Lotachukwu Ibe. Algorithmic bias: Investigating social biases in natural language processing models.

10. Tolga Bolukbasi, Kai-Wei Chang, James Y Zou, Venkatesh Saligrama, and Adam T Kalai. Man is to computer programmer as woman is to homemaker? debiasing word embeddings. *Advances in neural information processing systems*, 29, 2016.
11. Vipul Gupta, Pranav Narayanan Venkit, Shomir Wilson, and Rebecca J Passonneau. Survey on sociodemographic bias in natural language processing. *arXiv preprint* arXiv:2306.08158, 2023.
12. Thomas Davidson, Debasmita Bhattacharya, and Ingmar Weber. Racial bias in hate speech and abusive language detection datasets. In *Proceedings of the Third Workshop on Abusive Language Online*, pages 25–35, 2019.
13. Furkan Şahinuç, Eyup Halit Yilmaz, Cagri Toraman, and Aykut Koç. The effect of gender bias on hate speech detection. *Signal, Image and Video Processing*, 17(4):1591–1597, 2023.
14. Rachael Tatman. Gender and dialect bias in youtube's automatic captions. In *Proceedings of the first ACL workshop on ethics in natural language processing*, pages 53–59, 2017.
15. Alexandra Olteanu, Carlos Castillo, Fernando Diaz, and Emre Kıcıman. Social data: Biases, methodological pitfalls, and ethical boundaries. *Frontiers in big data*, 2:13, 2019.
16. Jing Qian, Hong Wang, Mai ElSherief, and Xifeng Yan. Lifelong learning of hate speech classification on social media. In *Proceedings of the 2021 Conference of the North American Chapter of the Association for Computational Linguistics: Human Language Technologies*, pages 2304–2314, 2021.
17. Syamsul Bahri, Elisa Betty Manullang, Putri Syah Nadillah Sihombing, and Kevin Enzo Eleazar. Language change in social media. *Randwick International of Social Science Journal*, 4(3):713–721, Jul. 2023. URL https://www.randwickresearch.com/index.php/rissj/article/view/745.
18. Reuben Binns. Fairness in machine learning: Lessons from political philosophy. In *Conference on fairness, accountability and transparency*, pages 149–159. PMLR, 2018.
19. Sabina Leonelli, Rebecca Lovell, Benedict W Wheeler, Lora Fleming, and Hywel Williams. From fair data to fair data use: Methodological data fairness in health-related social media research. *Big Data & Society*, 8(1):20539517211010310, 2021.
20. Dana Pessach and Erez Shmueli. Algorithmic fairness. In *Machine Learning for Data Science Handbook: Data Mining and Knowledge Discovery Handbook*, pages 867–886. Springer, 2023.
21. Gareth P Jones, James M Hickey, Pietro G Di Stefano, Charanpal Dhanjal, Laura C Stoddart, and Vlasios Vasileiou. Metrics and methods for a systematic comparison of fairness-aware machine learning algorithms. *arXiv preprint* arXiv:2010.03986, 2020.
22. Yan Zhou and Murat Kantarcioglu. On transparency of machine learning models: A position paper. In *AI for Social Good Workshop*, 2020.
23. Daniel Vela, Andrew Sharp, Richard Zhang, Trang Nguyen, An Hoang, and Oleg S Pianykh. Temporal quality degradation in ai models. *Scientific reports*, 12(1):11654, 2022.
24. Ed Shee. Bias vs Fairness vs Explainability in AI - Seldon — seldon.io. https://www.seldon.io/bias-vs-fairness-vs-explainability-in-ai. [Accessed 17-12-2023].
25. Ninareh Mehrabi, Fred Morstatter, Nripsuta Saxena, Kristina Lerman, and Aram Galstyan. A survey on bias and fairness in machine learning. *ACM computing surveys (CSUR)*, 54(6):1–35, 2021.
26. Daniel Borkan, Lucas Dixon, Jeffrey Sorensen, Nithum Thain, and Lucy Vasserman. Nuanced metrics for measuring unintended bias with real data for text classification. In *Companion Proceedings of The 2019 World Wide Web Conference*, pages 491–500, 2019.
27. Emilio Ferrara. Fairness and bias in artificial intelligence: A brief survey of sources, impacts, and mitigation strategies. *arXiv preprint* arXiv:2304.07683, 2023.
28. Simon Caton and Christian Haas. Fairness in machine learning: A survey. *ACM Computing Surveys*, 2020.
29. Marina Danilevsky, Kun Qian, Ranit Aharonov, Yannis Katsis, Ban Kawas, and Prithviraj Sen. A survey of the state of explainable ai for natural language processing. In *Proceedings of the 1st*

*Conference of the Asia-Pacific Chapter of the Association for Computational Linguistics and the 10th International Joint Conference on Natural Language Processing*, pages 447–459, 2020.
30. Binny Mathew, Punyajoy Saha, Seid Muhie Yimam, Chris Biemann, Pawan Goyal, and Animesh Mukherjee. Hatexplain: A benchmark dataset for explainable hate speech detection. In *Proceedings of the AAAI conference on artificial intelligence*, volume 35, pages 14867–14875, 2021.
31. Vijay Arya, Rachel KE Bellamy, Pin-Yu Chen, Amit Dhurandhar, Michael Hind, Samuel C Hoffman, Stephanie Houde, Q Vera Liao, Ronny Luss, Aleksandra Mojsilović, et al. One explanation does not fit all: A toolkit and taxonomy of ai explainability techniques. *arXiv preprint arXiv:1909.03012*, 2019.
32. Marco Tulio Ribeiro, Sameer Singh, and Carlos Guestrin. " why should i trust you?" explaining the predictions of any classifier. In *Proceedings of the 22nd ACM SIGKDD international conference on knowledge discovery and data mining*, pages 1135–1144, 2016.
33. Nikos Voskarides, Edgar Meij, Manos Tsagkias, Maarten De Rijke, and Wouter Weerkamp. Learning to explain entity relationships in knowledge graphs. In *Proceedings of the 53rd Annual Meeting of the Association for Computational Linguistics and the 7th International Joint Conference on Natural Language Processing (Volume 1: Long Papers)*, pages 564–574, 2015.
34. Fréderic Godin, Kris Demuynck, Joni Dambre, Wesley De Neve, and Thomas Demeester. Explaining character-aware neural networks for word-level prediction: Do they discover linguistic rules? In *Proceedings of the 2018 Conference on Empirical Methods in Natural Language Processing*, pages 3275–3284, 2018.
35. Qizhe Xie, Xuezhe Ma, Zihang Dai, and Eduard Hovy. An interpretable knowledge transfer model for knowledge base completion. In *Proceedings of the 55th Annual Meeting of the Association for Computational Linguistics (Volume 1: Long Papers)*, pages 950–962, 2017.
36. Scott M Lundberg and Su-In Lee. A unified approach to interpreting model predictions. *Advances in neural information processing systems*, 30, 2017.
37. LS Shapley. 17. a value for n-person games. contributions to the theory of games (am-28), 2016.
38. Edoardo Mosca, Ferenc Szigeti, Stella Tragianni, Daniel Gallagher, and Georg Groh. Shap-based explanation methods: a review for nlp interpretability. In *Proceedings of the 29th International Conference on Computational Linguistics*, pages 4593–4603, 2022.
39. Jay DeYoung, Sarthak Jain, Nazneen Fatema Rajani, Eric Lehman, Caiming Xiong, Richard Socher, and Byron C Wallace. Eraser: A benchmark to evaluate rationalized nlp models. In *Proceedings of the 58th Annual Meeting of the Association for Computational Linguistics*, pages 4443–4458, 2020.

# Mitigation

In the previous sections, the authors first explored how machine learning can be used to detect hate speech and later aimed to make the detection more explainable. This makes the moderation system is more reliable. In this section, the authors aim to understand what should be done once the authors detect hate speech. Several countries have established hate speech laws [1, 2] to deter harmful behaviour. In online settings, organizations like Facebook and Twitter opt to use methods such as blocking/suspension of messages/users to prohibit hate speech.[1] They have established policies [3] which define what is perceived as hate speech. Based on these policies moderators [4] are mandated to take down any hate speech present on these platforms.

While suspension/blocking can help the platform in reducing hate speech in the short term [5], the moderated users generally shift to other platforms [6] or create different accounts on the same platform. Hence, many researchers propose counterspeech, i.e., countering hate with appropriate speech as a solution that can work in the long term in reducing hate. Many of the social media companies have launched campaigns to promote counterspeech.[2] Nevertheless, similar to moderation, much of the onus of countering hate typically falls upon the moderators of the platforms and NGOs which are working to counter hate. Further the generated counter speech must adhere to certain guidelines, so that it can be effective in deescalation. Hence, the task of moderation is increasingly becoming much more challenging. One of the possible solutions is to generate counter speech in response to a hate speech, based on a text generation model like GPT, BART [7]. However, these kind of a system might face several challenges. The outputs might be unreliable making it difficult for the moderators to use these content without editing [8]. Another problem is the generation outputs might not be diverse enough in terms of their strategies [9]. This diversity in strategies

---

[1] https://tinyurl.com/tq8rp3o.

[2] https://counterspeech.fb.com/en/.

© The Author(s), under exclusive license to Springer Nature Switzerland AG 2026
P. Saha et al., *Online Hate Speech*, Synthesis Lectures on Human Language Technologies, https://doi.org/10.1007/978-3-031-86595-4_5

of counterspeech might be a desired quality of such a model as different communities might prefer different strategies during different forms of hate speech [10].

In this chapter, the authors will first discuss the different forms of mitigation techniques employed by the platforms for moderation of harmful content in terms of banning/suspension and understand their impact. The second part will discuss the concept of counterspeech, how the authors can add them to the moderation pipeline. The authors will also further discuss the usage of language models to generate counter speech.

## 5.1 Banning or Suspension

As mentioned above, banning or suspension is a common strategy employed by various social media platforms like Facebook and Twitter. Figure 5.1 provides an overview of the Facebook moderation system [11]. In this process, first, a post is identified as detrimental or not using some automatic moderation methods detailed in the previous sections. Then, the post is filtered for further processing and reviewed by human moderators. If the posts are still found to be violating policies, appropriate actions are taken, such as suspending the users or deleting the posts. Several factors determine whether something is violating the policies, including the context of the comment, cultural norms (e.g., language, country), the genre/style of the comment (e.g., humour, satire), or if a post was reproduced as a means of criticism and opposition. While there has been some movement towards proactive handling of such content, much of it remains reactive. Similar processes are followed on different platforms. One question that arises is whether such banning is effective or not. The different social media platforms like Facebook, and Twitter. Figure 5.1 gives an overview of the Facebook moderation system [11]. In this pipeline, first, a post is identified as detrimental or not using some automatic moderation methods as detailed in the previous sections. Then, the post is filtered for further processing and reviewed by the human moderators. If the posts are still found to be violating policies then appropriate actions are taken such as suspending the users or deleting the posts. There are various factors which help in deciding whether something is violating the policies. Usually, they are the context of the comment, cultural norms (e.g., language, country), the genre/style of the comment (e.g., humour, satire), or if a post was reproduced as a means of criticism and opposition. Although, there has been some movement towards proactive handling of such content, most of remains reactive. The similar pipeline is being followed at different platforms. One question that arises is whether such banning is effective or not.

### 5.1.1 Efficacy of Reddit Ban

In order to further understand the effect of banning, we explore the experiments done after one of the recent large scale bans on Reddit. On June 10, 2015, Reddit took action against

## 5.1 Banning or Suspension

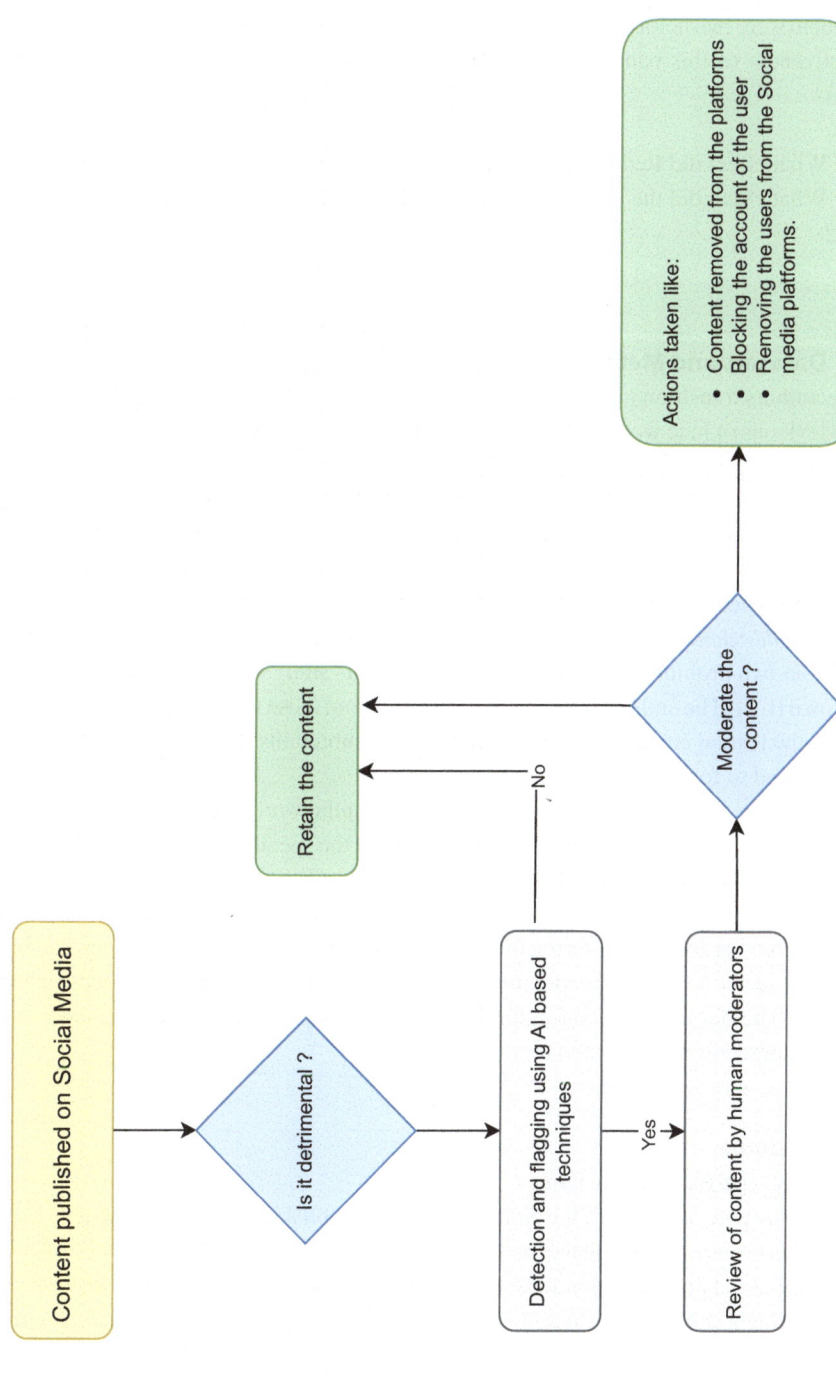

**Fig. 5.1** A figure to show the social moderation pipeline

some of the hateful subreddits, announcing that it would ban them under a new policy. Among them were two notorious subreddits: r/fatpeoplehate and r/CoonTown. In this section, the effectiveness of this ban is being studied [5]. They primarily focus on two research questions—

- RQ1: What effect did Reddit's ban have on the contributors to banned subreddits?
- RQ2: What effect did the ban have on subreddits that saw an influx of banned subreddit users?

#### 5.1.1.1 Datasets and Methods

Next, the authors transition to their dataset construction, and then describe the procedure to generate lexicons of hate words. The authors use the hate lexicons to perform their language analysis.

*Reddit 2015 corpus*: The authors collected a dataset of Reddit activity from 2015, including all publicly accessible submissions and comments. This dataset contains approximately 670 million posts (both submissions and comments) made between January and December 2015. They then extracted user and subreddit timelines from this data for further analysis.

*Banned subreddit data*: From the Reddit 2015 Corpus, the authors gather all the posts created in 2015 from two prohibited subreddits examined in this study: r/fatpeoplehate (FPH) and r/CoonTown (CT). The authors label the post datasets from these forums as DFPH and DCT. Extracting the textual content from posts within these subreddits allows them to create text collections used to form hate speech lexicons.

*Identification of hate speech*: As mentioned earlier, the authors rely on keyword-based methods, which are enhanced by tailoring the keywords to fit the specific community. Employing an automated technique for identifying keywords, the authors create lexicons specific to r/fatpeoplehate and r/CoonTown. This enables us to monitor if these lexicon words gain prominence in other forums following the ban. Subsequently, they manually review the lexicons generated automatically, pinpointing a selection of terms particularly geared towards hate speech. These refined lexicons, although more concise, provide greater accuracy and precision in targeting hate-oriented terms.

#### 5.1.1.2 Findings

**RQ1: User-level effects on the ban**

To examine any potential effects, the authors begin by establishing a set of comparable communities. This involves matching the treatment subreddits (r/fatpeoplehate and r/CoonTown) with control subreddits that might have faced bans, aligning the posting behavior of users from the treatment subreddits with users from control subreddits exhibiting similar patterns. The authors employ a difference-in-differences approach to analyze the differences before and after the bans between the treatment and control groups.

## 5.1 Banning or Suspension

Subsequently, users who posted at least five times were selected to mitigate the impact of random users. These chosen users underwent further matching based on their post-level activity within these groups. To achieve this, the authors utilized Mahalanobis Distance Matching (MDM), which calculates the distance between users based on three user characteristics (transformed to log scale).

- account age: days since user account was created
- karma: sum of scores on all comments made by the user
- total posts: total number of posts made pre-ban in 2015[9]

The user timeline examination occurred across two distinct periods: pre-ban and post-ban. Posts were then grouped into 10-day time windows for analysis. Upon analysis, it was observed that the usage of hate speech by users from the treatment groups notably decreased after the ban. The authors scrutinized a substantial volume of posts, specifically over 2.5 million from treatment CT and control CT users, and over 13 million from treatment FPH and control FPH users.

**RQ2: Community-level effects of the ban**
The authors employ an additional causal inference method, the Interrupted Time Series, to evaluate the ban's causal effects. The approach involves identifying subreddits that gained a significant number of users previously active in r/fatpeoplehate and r/CoonTown. Using this interrupted time series procedure, the authors compared the differences in hate speech before and after the ban within these invaded subreddits. The temporal fluctuations in hate speech usage within these affected subreddits are visually depicted in Fig. 3.2.

During their analysis, they examined over 9 million posts from subreddits invaded by migrants from r/CoonTown and more than 25 million posts from subreddits invaded by migrants from r/fatpeoplehate. Interestingly, hate speech usage within subreddits invaded by r/fatpeoplehate migrants remained relatively unchanged after the migration. In contrast, subreddits invaded by r/CoonTown migrants displayed an increase in hate speech usage post-migration. Notably, spikes in hate speech usage existed even before the ban, suggesting that these were not solely caused by the ban itself.

**Implications for online moderation** The decision by Reddit to ban r/fatpeoplehate and r/CoonTown, consequently scattering participants across other sections of the site, ultimately led to a decrease in overall hate speech usage on the platform. This outcome suggests that banning spaces where these groups gather can effectively curb hate speech. However, the decision of whether to ban groups for engaging in behaviors considered deviant by the site remains a challenging and unresolved question.

## 5.2 Counterspeech

Counterspeech serves as a direct response to harmful or hateful speech, aiming to undermine its impact. Platforms such as Facebook have established guidelines (like those at https://counterspeech.fb.com/en/) to assist the general public in countering hate speech online. However, crafting effective counterspeech poses significant challenges, as highlighted in research like Fumagalli [12].

Recently, various non-profit organizations have undertaken the responsibility of combating online hate, as seen with initiatives like "wecounterhate" platform.[3] While this approach of crowdsourcing counterspeech efforts can be effective, it places a considerable mental burden on NGO operators due to the sheer volume of hateful content generated daily, as discussed in studies such as Vidgen et al. [13].

The paper [14] outlines eight strategies for countering hateful messages online.However, in this book, following the previous work [10], we've decided to focus on seven strategies, modifying the 'Tone' category to solely include 'Positive Tone' as a valid method of countering hate speech. Another category, 'Hostile,' is acknowledged but discouraged due to its potential to escalate hateful conversations. It's important to note that a single counterspeech instance can incorporate multiple strategies. Additionally, while these strategies cover a range of approaches, they might not encompass all possible types of counterspeech strategies, leaving room for other potential methods not included in this specific classification.

**Presenting facts to correct misstatements or mis-perceptions**: In this approach, the person countering the speech aims to persuade by correcting inaccuracies or misstatements. For instance, in theirdataset, there's an example of this type of counterspeech directed towards the LGBT community. The response was: *"Actually homosexuality is natural. Nearly all known species of animal have their gay communities. Whether it be a lion or a whale, they have or had (if they are endangered) a gay community. Also marriage is an unnatural act. Although there are some species that do have longer relationships with a partner most known do not"* This comment was made in response to an interview video where the interviewee claimed that homosexuality is unnatural, detrimental, and harmful to society.

**Pointing out hypocrisy or contradictions**: In this strategy, the person speaking against hate points out the inconsistency or hypocrisy in the individual's hateful statements. To refute the accusation, the individual might explain or justify their past behavior, or if they're open to persuasion, commit to avoiding such conflicting behavior in the future [15]. An example of this type of response directed at the LGBT community from their dataset is: *"The 'US Pastor' refuses to accept gays citing the Bible's prohibition. Yet, he conveniently overlooks numerous other biblical prohibitions—eating shrimp or pork, touching the skin of a dead pig (Football), mixing fabrics, wearing torn clothes, getting tattoos, planting two crops in one field, working on the Holy Day (Saturday or Sunday)... according to the Bible, these actions should result in eternal punishment, yet they're ignored. But when it comes to*

---

[3] https://wecounterhate.com/.

*loving someone considered 'wrong' (gays), suddenly it's unacceptable! This cherry-picking of the Bible only to support bigotry exposes hypocrisy. YOU'RE A HYPOCRITE.*" This comment was a response to an interview video where the interviewee promoted hatred against homosexual people

**Warning of offline or online consequences**: This strategy involves the counterspeaker cautioning the original speaker about the potential consequences of their actions. Sometimes, this can lead the initial speaker of hate speech to reconsider or retract their original opinion. An example of this type of response directed at the LGBT community from their dataset is: ""*I'm not gay, but regardless of someone's sexual orientation, assaulting anyone is wrong. This preacher should face legal consequences for inciting violence and sexual harassment!*"". This comment was in response to a video where a preacher advised people to physically harm children if they identified as gay.

**Affiliation**: Affiliation in this context refers to establishing, maintaining, or restoring positive relationships with others. Research suggests that people are more inclined to trust and credit the counterspeech of those they feel affiliated with, as they view ingroup members as more trustworthy, honest, loyal, cooperative, and valuable compared to outgroup members.

In their dataset, counterspeakers who employ the strategy of affiliation receive the highest number of likes for their comments among all the counterspeech types. An example of this type of response directed at the LGBT community from their dataset is: "*Hey, I'm Christian and I'm gay. This guy's completely wrong. It's time to stop justifying and start accepting. I know where my heart and soul belong—to God, the creator of heaven and earth. The authors all exist in His realm of understanding, so it's high time the authors started accepting one another. That's all.*" This comment was in response to an interview video where the interviewee promoted hate against homosexual people.

**Denouncing hateful or dangerous speech**: In this strategy, the counterspeakers denounce the message as being hateful. This strategy can help the counterspeakers in reducing the impact of the hate message. An example of this type of counterspeech toward *Jews* community from their dataset is as follows: "*please take this down youtube. this is hate speech.*". This comment was in response to a video in which a preacher is advocating hatred and killing of Jewish people.

**Humor and sarcasm**: Humor is indeed a potent tool wielded by counterspeakers against hate speech. It has the ability to defuse tensions and draw attention to the issue at hand. In online settings, humor can soften hostility, provide support to other speakers, and foster social unity. Sarcasm is often a form of humor used in these situations, as seen in this example directed at the LGBT community in their dataset: "*HAHAHAHAHAHAHAH...oh you were serious. That's even funnier :)*". This comment was a response to a video where a preacher advocated hate and violence towards homosexual people.

**Positive tone**: The counterspeaker uses a wide variety of tones to respond to hate speech. In this strategy, the authors consider different forms of speech such as empathic, kind, polite, or civil. Increasing empathy with members of opposite groups counteracts incitement [16]. An example of this type of counterspeech toward *Jews* community from their dataset is as follows: *"I am a Christian, and I believe we're to love everyone!! No matter age, race, religion, sex, size, disorder...whatever!! I LOVE PEOPLE!! the authors are not going to go anywhere as a country if the authors don't put God first in our lives, and treat EVERYONE with respect"*. This comment was in response to a video in which a preacher is advocating hatred and killing of Jewish people.

**Hostile language**: Absolutely, employing different tones in responding to hate speech is crucial. This strategy involves using various forms of speech, including empathetic, kind, polite, or civil tones. Research indicates that increasing empathy towards members of opposing groups can counteract incitement. While the original definition of "Tone" encompassed hostile counterspeech, we've chosen to distinguish "Hostile language" as a separate type of counterspeech. An example of a response directed at the Jewish community from their dataset is: *"I am a Christian, and I believe we're to love everyone!! No matter age, race, religion, sex, size, disorder...whatever!! I LOVE PEOPLE!! the authors are not going to go anywhere as a country if the authors don't put God first in our lives, and treat EVERYONE with respect"*. This comment was in response to a video where a preacher advocated hatred and violence towards Jewish people.

## 5.3 Generation of Counterspeech

The generation of counterspeech essentially becomes a task in which there is a hate speech as an input, which passes through a system and the system produces counterspeech. This system for our case is essentially and language model. In very simple terms, a language model is a probabilistic model of a natural language. Early language models relied on statistical approaches and n-gram models to predict the likelihood of a word given its context. These models had limitations in understanding complex language structures . Rule-based systems incorporated linguistic rules and grammatical structures to analyze and generate text. However, they struggled with handling ambiguity and varied language patterns. With advancements in computational power and neural network architectures, recurrent neural networks (RNNs) and long short-term memory networks (LSTMs) began to model sequential data more effectively. These models showed improvements in language understanding and generation. Finally, The introduction of the Transformer architecture, particularly the Attention mechanism, revolutionized language modeling. Models like "BERT" (Bidirectional Encoder Representations from Transformers) and "GPT" (Generative Pre-trained Transformer) emerged, enabling bidirectional context understanding and achieving state-of-the-art performance on various natural language processing tasks.

## 5.3.1 Generation Models

the authors primarily work on the latest transformer models for the generation of counter-speech. There are three types of transformer models as noted below:

### 5.3.1.1 Autoencoder Models

Autoencoding models play a significant role in language understanding and classification tasks by focusing on the encoding aspect. These models aim to create meaningful representations of input data through their encoded format. They're designed for comprehension tasks, capturing crucial information from input sequences.

Autoencoding models typically use bidirectional training, considering both forward and backward contexts in the input sequences. This bidirectional approach helps them better understand relationships and connections within the text. During training, autoencoders use masking techniques intentionally, concealing or distorting specific parts of the input. This approach encourages the model to learn robust features and develop the ability to reconstruct and identify missing or distorted information. By doing so, they become adept at grasping the essential elements of the input data and understanding how different parts relate to each other

### 5.3.1.2 Autoregressive Models

Autoregressive models, like GPT (Generative Pre-trained Transformer), operate by predicting subsequent tokens based on prior ones in an iterative manner. These models heavily utilize the decoder component of the transformer architecture and employ probabilistic inference to generate text. Unlike sequence-to-sequence models, autoregressive models don't require an explicit input sequence and excel at text generation tasks.

Although autoregression isn't unique to Transformers and finds applications in domains like Time Series Forecasting, AR Transformers stand out due to their adaptability through fine-tuning. However even with very good performance, these models also generate text without a deep understanding of the underlying context. This limitation becomes apparent in text generation where the output, while structurally sound, may lack reliability, leading to inaccuracies or generating content that doesn't align with reality. This issue is particularly noticeable in simpler tasks like text summarization.

It's crucial to acknowledge that relying solely on GPT-based summary tools might result in misleading or incorrect information. Fortunately, alternative Transformer architectures are emerging that offer more comprehensive language understanding and improved accuracy for tasks requiring deeper linguistic comprehension.

### 5.3.1.3 Seq2Seq Models

Seq2Seq model or Sequence-to-Sequence model, is a machine learning architecture designed for tasks involving sequential data. It takes an input sequence, processes it, and generates an output sequence. The architecture consists of two fundamental components: an encoder and a decoder. Now these encoder and decoder can be represented by transformer models as well. Initially, tailored for translation purposes, these models excelled in mapping sequences between different languages. This is because these architectures are specifically designed to handle sequence mapping. The separate utilization of encoder and decoder blocks presents a significant advantage: scalability across newer languages becomes relatively straightforward.

## 5.3.2 Decoding Methods

Different decoding methods utilise different ways to generate text from these language models. This section provides an overview of diverse decoding techniques: Greedy search, Beam search, Sampling, Top-K sampling, and Top-P (nucleus) sampling. The authors further use an example (as shown in Fig. 5.2) to highlight the decoding strategies.

*Greedy Search*: Greedy search selects from the language model the word with the highest probability as the next word. Suppose the language model is predicting a possible continuation to a sentence starting with "The", choosing from the words "red" and "cat" with their associated probabilities. Greedy search always chooses the word with the highest probability, which is "cat".

*Beam search*: Beam search addresses this problem by keeping the most likely hypotheses (a.k.a. beams) at each time step and eventually choosing the hypothesis that has the overall highest probability. Suppose the authors are performing a beam search with two beams. At the first timestep, beam search would consider both the words "red" and "cat", along with their probabilities. At the second timestep, beam search will consider all the possible

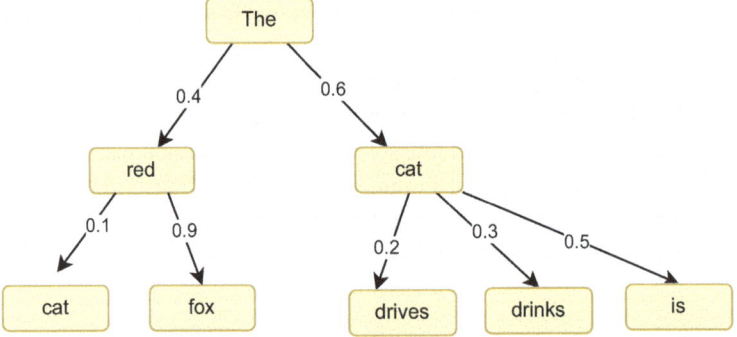

**Fig. 5.2** Example of word probabilities as predicted by the language models

## 5.3 Generation of Counterspeech

continuations of the two beams and keep only the two of them with the highest probabilities. Thus, the two beams at the next timestep will be "The red fox" and "The cat is".

*Sampling*: Sampling means randomly picking the next word according to the conditional word probability distribution extracted from the language model. As a consequence, with this decoding method text generation is not deterministic. Using directly the probabilities extracted from the language models often leads to incoherent text. A trick is to make the probability distribution sharper (as in increasing the likelihood of high-probability words and decreasing the likelihood of low-probability words) by applying a softmax over the probability distribution and varying its temperature parameter to make it sharper or smoother. With this trick, the output is generally more coherent.

*$Top_k$ sampling*: In $Top_k$ sampling, the $k$-most likely next words are filtered and then the next predicted word will be sampled among these K words only. The produced text is often more human-sounding than the text produced with the other methods seen so far. One concern though with $Top_k$ sampling is that it does not dynamically adapt the number of words that are filtered from the next word probability distribution, as $k$ is fixed. As a consequence, very unlikely words may be selected among these $k$ words if the next word probability distribution is very sharp.

*$Top_p$ (nucleus) sampling*: $Top_p$ sampling (or nucleus sampling) chooses from the smallest possible set of words whose cumulative probability exceeds the probability $p$. This way, the number of words in the set can dynamically increase and decrease according to the next word probability distribution.

### 5.3.3 Evaluation Metrics

After generation, the authors have to evaluate the generated counterspeech in order to evaluate various aspects of the same. There are three different forms of evaluation metrics—**referential** metrics, **aspect** metrics and **human evaluation** metrics.

#### 5.3.3.1 Referential Metrics

These metrics are measured using comparing the generated texts and the ground truth response. These are the metrics used—

- *BLEU*—The metric used in comparing a candidate translation to one or more reference translations. And the output lies in the range of 0–1, where a score closer to 1 indicates good quality translations.

$$\text{BLEU} = \min\left(1, \frac{\text{output-length}}{\text{reference-length}}\right) \left(\prod_{i=1}^{4} \text{precision}_i\right)^{\frac{1}{4}} \quad (5.1)$$

- *meteor*—The measurement relies on the harmonic average of how accurately individual words are identified and remembered, favoring remembering over precise identification. Additionally, it includes distinctive elements like analyzing word roots and finding synonymous terms, in addition to the usual exact word comparison, setting it apart from other metrics.
- *bleurt*—bleurt, or Bilingual Evaluation Understudy with Representations and Transformers, is a specialized metric for evaluating machine translation quality. It compares machine-generated translations with human references, using transformer-based models to measure linguistic accuracy, fluency, and semantic alignment. This nuanced evaluation, adaptable to various languages and domains, offers a refined assessment aligned closely with human judgment, distinguishing it from simpler metrics like BLEU or METEOR.
- *bertscore*—bertscore is an evaluation metric used to assess the quality of machine-generated text, particularly in natural language generation tasks like machine translation and summarization. It measures the similarity between the generated text and reference text using BERT embeddings, a type of language representation model. BERTScore calculates a similarity score by considering contextual embeddings of words, offering a more nuanced evaluation compared to traditional metrics like BLEU or ROUGE. Its ability to capture semantic meaning and context enables a more accurate assessment of the quality of generated text, making it a valuable tool for evaluating language generation models.
- *Diversity*—The diversity [17] of the given set of generated sentences $s$ is defined in Eq. 5.2. $\psi$ is the Jaccard similarity function.

$$diversity(s) = (1/|s|) * \sum_i 1 - max((\psi(s_i, s_j))_{j=1}^{j=|s|, j!=i} \qquad (5.2)$$

- *Novelty*—The novelty of the generated outputs can be calculated the novelty [17] using Eq. 5.3 where $c$ is the sentence set of training corpus and $\psi$ is the Jaccard similarity function. This is done to identify if the samples directly copied from the training samples or the model is trying to generate of its own.

$$novelty(s) = (1/|s|) * \sum_i 1 - max((\psi(s_i, c_j))_{j=1}^{j=|c|} \qquad (5.3)$$

### 5.3.3.2 Aspect Metrics

In the previous section, the authors saw that there are different forms of referential metrics which use the ground truth counterspeech. In practice, there might be different aspects which might not be present in the ground truth counterspeech but are desirable in counter speech in general. One common way to do this will be using a third party classifier which identify that particular aspect. Here the authors represent such possible aspect identifiers.

## 5.3 Generation of Counterspeech

*Engagement prediction*: One of the important properties of the counterspeech will be to increase the amount off positive or hope in the audience. For this measuring the engagement is very important [18]. One may use the DialogRPT model [19] to predict the human feedback of the counterspeech generated in terms of the following metrics—*width*: the number of direct replies to the given reply, *depth*: the maximum length of dialog after this turn, and *updown*: the number of upvotes minus the number of downvotes. To calculate the engagement metric, the authors pass the `hate speech-counterspeech` pair to the model which provides a score between 0 and 1 representing the engagement in terms of upvotes/width/height. This will denote the engagement probability of that metric for the given counterspeech.

*Quality-metrics*: Further, one can employ various third-party classifiers to evaluate the quality of the generated responses. To calculate the scores, the authors pass the generated counterspeech through the model and get the logit scores which are passed through a softmax layer. The metrics used for evaluation are listed below.

- *Counterspeech*: In order to evaluate the counterspeech quality of the generated responses, the authors can use a `bert-base-uncased` model trained on a dataset which contains COUNTERSPEECH and COUNTERSPEECH and use one of the language models like `bert-base-uncased` model. Given this model, the authors pass each generated response through the classifier to predict a confidence score which denotes the quality of the counterspeech. This is a reference free metric for counter speech quality measurment.
- *Counter-argument*: In order to evaluate the counter argument characteristic of the generated response, the authors can use a `bert-base-uncased` model trained on the counter argument dataset which contains counter argument pairs. Given this model, the authors pass each the hate speech and the generated response through the classifier to predict a confidence score which would denote the counter argument quality.
- *Toxicity*: the authors can use the HateXplain model [20] trained on two classes—toxic and non-toxic.[4]

### 5.3.4 Human Evaluation

Finally, since this is a subjective generative task, the authors additionally need humans to evaluate the quality of the generated texts [21]. Following metrics are utilised in order to evaluate the same—

---

[4] https://huggingface.co/Hate-speech-CNERG/bert-base-uncased-hatexplain-rationale-two.

- *Suitableness (SUI)*—This metric measures how suitable a counterspeech is to the hate speech in terms of semantic relatedness and in terms of adherence to counterspeech guidelines.
- *Grammaticality (GRM)*—This metric measures how grammatically correct the particular counterspeech according to the annotator.
- *Specificity (SPE)*—This metric measures how specific are the arguments brought by the counterspeech in response to the hatespeech.
- *Choose-or-not (CHO)*—This metric measures whether the annotators would select the counterspeech to post-edit and use it in a real case scenario as in the setup presented by this paper [21].
- *Is-best (BEST)*—The metric defines whether the counterspeech is the absolute best among the ones generated.

The first three metrics are measured using a 5 points Likert-scale whereas for the other two are binary rating (0/1). Choose-or-not allows for multiple counterspeech to be selected for the same hate speech, while only one counterspeech can be selected for is-best for each hate speech.

## 5.4 Comparison Across Generation Strategies

In the previous section, the authors discussed various models and decoding strategies for generation of counterspeech. In this section, the authors try to empirically explore different models and decoding strategies [21]. To compare different models and decoding strategies, the authors rely on good quality counter speech dataset [22]. This dataset further provides diverse targets as noted in Appendix. The dataset was partitioned into training, validation, and test sets with the ratio: 8 : 1 : 1 (i. e. 4003,500 and 500 pairs), ensuring that all sets share the same target distribution, and no repetition of HS across the sets is allowed. The authors experiment with 5 Transformer based language models representing the main categories of the model mechanisms: **autoregressive**, **autoencoder**, and **seq2seq**.

1. *BERT*—The Bidirectional Encoder Representations from Transformers was introduced by this work [23]. It is a bidirectional autoencoder that can be adapted to text generation [24]. The authors warmstarted an encoder-decoder model using BERT checkpoints similar to the BERT2BERT model defined by this work [25].
2. *GPT-2*—The Generative Pre-trained Transformer 2 is an autoregressive model built for text generation [26].
3. *DialoGPT*—The Dialogue Generative Pretrained Transformer is the extension of GPT-2 specifically created for conversational response generation [27].

## 5.4 Comparison Across Generation Strategies

4. *BART*—BART is a denoising autoencoder for pretraining seq2seq models [28]. The encoder-decoder architecture of BART is composed of a bidirectional encoder and an autoregressive decoder.
5. *T5*—T5. The Text-to-Text Transfer Transformer proposed by the work [29] is a seq2seq model with an encoder-decoder Transformer architecture.

All the other models except BERT was fine-tuned directly for the generation task. the authors utilize 4 decoding mechanisms: a deterministic (Beam Search) and three stochastic (Top-$k$, Top-$p$, and a combination of the two).

1. *Beam Search (BS)*: The Beam Search algorithm is designed to pick the most-likely sequence [30, 31].
2. *$Top_k$*: The sampling procedure proposed by the work [32] selects a random word from the $k$ most probable ones, at each time step.
3. *$Top_p$*: Also known as Nucleus Sampling, the parameter $p$ indicates the total probability for the pooled candidates, at each time step [33].
4. *Combining $Top_{pk}$*: At decoding stage, it is possible to combine the parameters $p$ and $k$. This is a Nucleus Sampling constrained to the Top—$k$ most probable words.

Some of the default parameters are: Beam-Search with 5 beams and repetition penalty = 2; $Top_k$ with $k = 40$; $Top_p$ with $p = 0.92$; $Top_{pk}$ with $k = 40$ and $p = 0.92$. This was primarily used for doing these comparisons.

### 5.4.1 Best Model

The outcomes from comparing various models' $Best_{LM}$ generations are detailed in Table 5.1. In terms of overlap and diversity metrics, DialoGPT consistently achieves the top or second-best scores across all metrics, except for novelty, where it still attains a high score (0.643), closely trailing the best performance (0.655). T5 also demonstrates strong performance, especially in ROUGE, BLEU-1, and novelty.

In contrast, BART emerges as the leading model in human evaluation metrics, except for specificity. However, it displays weak performance in diversity metrics, suggesting a tendency to generate grammatically correct yet overly generic responses. BERT, on the other hand, registers the lowest scores across overlap and diversity metrics except for novelty. Despite achieving superior syntactic metrics, BERT's output appears more complex yet highly repetitive. This combination adversely impacts the clarity of its output, resulting in lower human evaluation scores, notably in grammaticality (4.2, while other models score above 4.6). The poor grammaticality might also account for the syntactic scores, as the

**Table 5.1** Results of the overlap and diversity metrics are calculated on the $Best_{LM}$ generations while the toxicity, the syntactic metrics and the human evaluation are calculated on the corresponding subset

| | Overlap | | | | Diversity | | Toxicity | Syntactic metrics | | | Human evaluation | | | | |
|---|---|---|---|---|---|---|---|---|---|---|---|---|---|---|---|
| | ROU | B1 | B3 | B4 | RR | NOV | – | ASD | MSD | NST | SUI | SPE | GRM | CHO | BEST |
| BART | 0.268 | 0.277 | 0.085 | **0.051** | 20.722 | 0.560 | 0.420 | 4.311 | 4.965 | 1.740 | **3.790** | 2.552 | **4.937** | **0.840** | **0.272** |
| BERT | 0.237 | 0.277 | 0.073 | 0.037 | 24.747 | 0.605 | 0.406 | **5.008** | **6.160** | **2.280** | 3.135 | 2.647 | 4.247 | 0.717 | 0.122 |
| T5 | **0.274** | **0.302** | 0.083 | 0.042 | 8.548 | **0.655** | 0.359 | **4.692** | 5.325 | 1.715 | 2.872 | 2.402 | 4.680 | 0.642 | 0.090 |
| DialoGPT | **0.273** | **0.304** | **0.093** | **0.052** | **8.248** | 0.643 | **0.343** | 4.677 | 5.575 | 1.895 | 3.392 | **2.755** | **4.880** | 0.767 | 0.245 |
| GPT-2 | 0.264 | 0.297 | **0.088** | 0.050 | **7.736** | 0.653 | **0.342** | 4.584 | 5.595 | 2.240 | 3.555 | **2.880** | 4.867 | **0.795** | **0.270** |

## 5.4 Comparison Across Generation Strategies

spaCy dependency parser wasn't trained to handle ungrammatical text, potentially inflating the ASD and MSD scores.

GPT-2 demonstrates overall competitive results across multiple metric categories. It secures the second-highest novelty score (0.653) and the best RR (7.736). Additionally, it performs second-best in BLEU-3, maximum syntactic depth, and number of sentences. Moreover, GPT-2 achieves top results in toxicity and specificity (2.880), indicating its capability to generate complex, suitable, focused, and diverse counter narratives (CNs).

### 5.4.2 Best Decoding Method

The outcomes derived from analyzing the BestD output are showcased in Table 5.2. Among the decoding mechanisms, $Top_k$ emerges as the top performer, excelling in diversity metrics, BLEU-3, and BLEU-4. It also leads in specificity, maximum syntactic depth, and number of sentences, while securing the second-best positions in average syntactic depth and toxicity. Other stochastic decoding mechanisms also exhibit strong performance. $Top_p$ demonstrates competitive results across diversity, overlap metrics, and secures the second-best position in specificity, alongside commendable achievements in syntactic metrics. $Top_{pk}$ performs well in overlap metrics, achieving second-highest scores in most human evaluation metrics, lowest in toxicity, and a reasonable specificity score. Conversely, BS doesn't yield notably strong outcomes except for the ROUGE score. Although it excels in human evaluation metrics, it sacrifices specificity and diversity. Further analysis revealed that this was due to the deterministic nature of BS, favoring the most probable sequences—resulting in vague and repetitive outputs.

### 5.4.3 Best Model-Decoding Combination

Here, the authors briefly analyze the evaluation outcomes derived from the BestLM+D generations. Specifically, autoregressive models like GPT-2 and DialoGPT exhibit similar behaviors, especially concerning similar decoding mechanisms. For instance, BS yields the best results across nearly all overlap metrics but performs poorly in diversity metrics. Conversely, with other models, stochastic decoding mechanisms prove to be superior for overlap metrics. Across various cases, the authors notice that stochastic decoding mechanisms generally outperform in syntactic, diversity metrics, and toxicity. However, in terms of human evaluation metrics, BS tends to excel, except in specificity.

**Table 5.2** The results for the overlap and diversity metrics are calculated on the BestD generations: for each decoding mechanism, there are 2500 CNs. The remaining metrics are calculated on a subset of 1000 CNs; the distribution of which is shown in the column "n"

| | Overlap | | | | Diversity | | Toxicity | Syntactic metrics | | | Human evaluation | | | | | n |
|---|---|---|---|---|---|---|---|---|---|---|---|---|---|---|---|---|
| | ROU | B1 | B3 | B4 | RR | NOV | | ASD | MSD | NST | SUI | SPE | GRM | CHO | BEST | |
| BS | 0.287 | 0.299 | 0.096 | 0.059 | 21.579 | 0.5 | 0.398 | 4.415 | 5.1 | 1. | 3.936 | 2.497 | 25 | 0. | 0.222 | %18.7 |
| Top$_{pk}$ | 0.287 | 0.320 | 0.106 | 0.059 | 11.404 | 0.639 | 0.352 | 4.676 | 5.488 | 1.932 | 3.324 | 2.647 | 4.688 | 0.764 | 0.212 | %29.3 |
| Top$_k$ | 0.282 | 0.314 | 0.106 | 0.060 | 10.076 | 0.6 | 0.374 | 4.704 | 5.756 | 2.133 | 3.155 | 2.716 | 4.659 | 0.716 | 0.183 | %27.1 |
| Top$_p$ | 0.285 | 0.319 | 0.105 | 0.060 | 11.270 | 0.640 | 0.381 | 4.753 | 5.671 | 2.068 | 3.149 | 2.687 | 4.681 | 0.723 | 0.189 | %24.9 |

## 5.5 Generate, Prune and Select

In the previous section, the authors discussed strategies to generate counterspeech and further tried to understand which combination of models and decoding methods are actually improving the performance. In this and the next sections, the authors look into how the authors can improve the performance of such models. The first proposed method is Generate, Prune and Select method [34], which uses additional tools to make the quality of the generated texts better.

### 5.5.1 Proposed Model

As in any counterspeech generation task, the authors assume access to a corpus of labeled pairs of conversations $\mathcal{D} = \{(x_1, y_1), (x_2, y_2), \ldots, (x_n, y_n)\}$, where $x_i$ is a hate speech and $y_i$ is the appropriate counterspeech as decided by experts or by crowd sourcing. The goal is to learn a model that takes as input a hate speech $x$ and outputs a counterspeech $y$. The authors present an overview of the model in Fig. 5.3 and describe each module in detail below.

#### 5.5.1.1 Candidate Generation
The primary objective of this module is to expand the pool of potential counterspeech options. The authors start by gathering all available counterspeech instances, represented as $Y = [y_1, y_2, \ldots, y_n]$, from the training dataset. Then, they employ a generative model, specifically an RNN-based variational autoencoder (referencing Bowman et al.), designed to utilize global distributed latent representations of sentences. This model aims to augment the counterspeech pool.

Both the encoder and decoder architectures consist of two layers, each comprising 512 nodes. To ensure robust training, the authors integrate two highway network layers (citing Srivastava et al.). Similar to other generative models, its objective is to maximize the lower bound likelihood $\mathcal{L}$ in generating the training data $Y$.

$$\mathcal{L} = -KL\left(q_\theta(z \mid y) \| p(z)\right) + \mathbb{E}_{q_\theta(z \mid y)}\left[\log p_\theta(y \mid z)\right]$$

In this setup, $\theta$ encompasses all parameters within the generative model. The variable $z$ represents a latent factor following a Gaussian distribution with a diagonal covariance matrix. Here, $p$ denotes the prior distribution, while $q$ denotes the posterior distribution, and $KL$ refers to the KL-divergence (citing KL divergence).

Throughout the training process, the authors implement the KL annealing technique (inspired by Bowman et al.) to mitigate the stable equilibrium issue, specifically addressing scenarios where the first term of the likelihood function, $KL\left(q_\theta(z \mid y) | p(z)\right)$, tends towards

**Fig. 5.3** Overview of generate prune select architecture [34]. The red ovals correspond to the individual modules

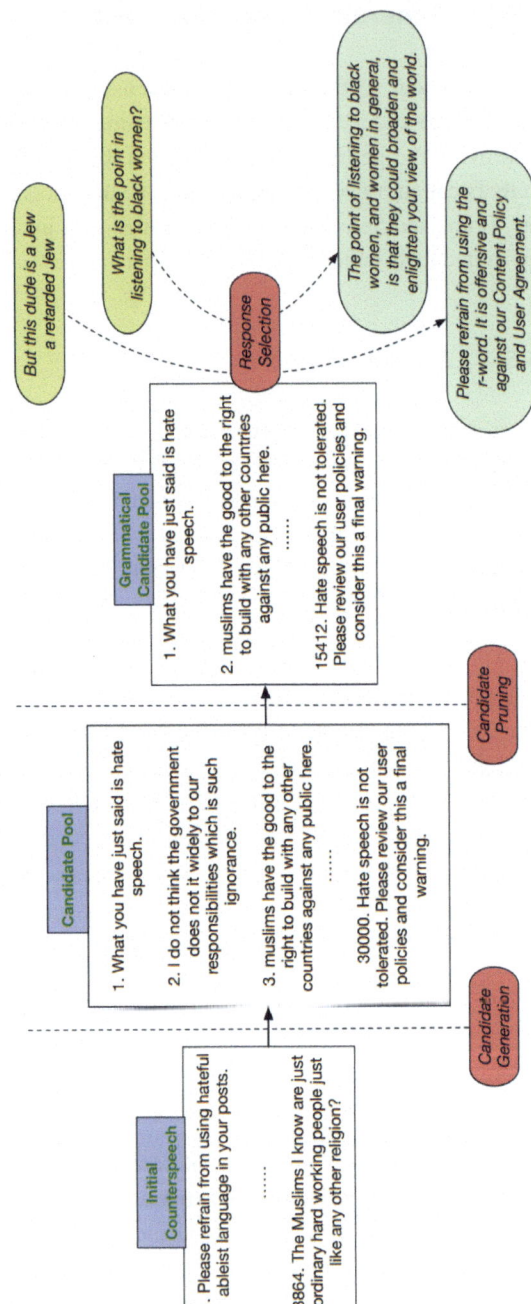

zero undesirably. Once training is complete, candidate generation involves decoding noise $\epsilon$ obtained from a standard Gaussian distribution (i.e., $\epsilon \sim \mathcal{N}(0, 1)$).

It's worth noting that this architecture can readily be substituted with other contemporary transformer architectures like GPT2, DialoGPT, or T5. This flexibility allows for similar experiments to be conducted using these alternate architectures.

### 5.5.1.2 Candidate Pruning

Though candidates generated by such an RNN-based variational autoencoder are diverse, they are not always grammatical as pointed out by this work [35]. Therefore, in this module, the authors prune the candidate list and retain only the grammatical ones. Toward this, the authors train a grammatically classifier on the corpus of linguistic acceptability (CoLA), a dataset with 10,657 English sentences labeled as grammatical or ungrammatical from linguistics publications. The authors select BERT as the classification model, and fine-tune it on the CoLA dataset. The choice of BERT is to best capture both the syntactic and the contextual information, and the authors select the 'bert-base-cased' model for its better computational efficiency.

### 5.5.1.3 Response Selection

The authors now have a collection of diverse and grammatical counterspeech responses. Finally, the authors aim to select the most relevant response to a given hate speech instance.

Taking into consideration the limited training instances that are realistically available, and inspired by the recent success of pretrained models, the authors innovate on a pretrained response selection model for task-oriented dialogue systems [36] and perform fine-tuning on their dataset. The paper [36] proposed two response selection methods, but the authors find that neither of them is well-suited for their task.

- *Train a response selection classifier with the negative sampling technique*: It relies on randomly drawing other candidates from the candidate pool as negative examples. However, in their task, one hate speech instance usually has multiple appropriate counterspeech instances. For example, given the hate speech "I am done with Islam and ISIS. All Muslims should be sent to their homeland. Britain will be better without their violence and ideology.", there are many other instances that can work as quality counterspeech, such as "You cannot blame all people for the actions of a few. Banning something altogether will not solve anything." or "Does prohibition of anything ever work? I thought religious tolerance was one of our 'British values'?". Therefore, many wrongly chosen negative examples may negatively impact the inductive bias of the response selection classifier.
- *Select by cosine similarity*: the authors point out that the embeddings of the input (hate speech) and the responses (counterspeech candidates) do not share the same latent vector

space and therefore, the learned embeddings and their cosine similarities may not fully serve the purpose of relating the response to the input.

Therefore, instead of adopting the two available methods directly, the authors improve on the second one by fusing the latent spaces of the input and the responses, inspired by this work [37]. Specifically, the authors propose to learn a linear embedding mapping from the latent space of the responses to the latent space of the input, and then select the best response by cosine similarity.

Mathematically, the authors use $e_x$ to denote the input embedding and $e_y$ to denote the response embedding. The authors aim to learn a linear mapping from $e_y$ to $e'_y$, where $e'_y = (W + BI) \cdot e_y$, $W$ and $B$ are learnable parameters, and $I$ is an identity matrix. The authors learn the mapping such that the sum of the cosine similarities between $e_x$ and $e'_y$ for the training data is maximized. By way of this transformation, $e'_y$ now maps the vector space of the responses to that of the input, and thus allows the pretrained model to effectively utilize the discriminative power of the sentence embeddings.

### 5.5.2 Experiments

*Datasets*: To evaluate this proposed model, the authors used three datasets—CONAN, Gab and Reddit. This work [9] has fully-labeled hate speech intervention datasets collected from Reddit and Gab, comprising 5,257 and 14,614 hate speech instances respectively. The authors use the filtered conversation setting in Qian et al. [9], which includes the posts labeled as hate speech only and discards other non-hateful conversations. Besides, the authors use the English language portion of the CONAN dataset [38], which contains counterspeech for 408 hate speech instances, written by experts trained on countering hatred. The Reddit, Gab and CONAN datasets have on average 2.66, 2.86 and 9.47 ground truth counterspeech for each hate speech respectively.

*Training data*: Since each hate speech can have multiple ground truth counterspeech, the authors follow the work [9] to dis-aggregate the counterspeech and construct a pair (hate speech, counterspeech) for each of the ground truth counterspeech in each dataset. Given a counterspeech dataset, the authors randomly choose 70% (hate speech, counterspeech) pairs for model training, 15% for cross validation and the rest 15% for testing.

*Baselines*: the authors compare their proposed models with four competitive baselines.

1. *Seq2seq* This method [39] is a widely used neural model for language generation. The authors use 2 bidirectional Gated Recurrent Unit (GRU) layers for the encoder and 2 GRU layers followed by a 3-layer neural network as the decoder.
2. *Maximum Mutual Information (MMI)* [40] is a diversity-promoting approach for neural conversation models. The authors implement the MMI-bidi model [40] and adopt incremental learning [41] to facilitate robust training.

## 5.5 Generate, Prune and Select

3. *SpaceFusion* [37] optimizes both diversity and relevance by introducing a fused latent space, where the direction and distance from the predicted response vector roughly match the relevance and diversity, respectively. The authors align the direction parameter with the ground truth counterspeech. To better exercise the diversity power, the authors randomly choose the distance parameter at each time of generation.
4. *BART* [28] is the state-of the-art pre-trained sequence-to-sequence model for language generation. It has a standard Transformers-based neural machine translation architecture which can be seen as generalizing BERT [23], GPT [26] and many pretraining schemes. The authors finetune the BART model on their training data.

*Evaluation*: the authors evaluate all model outputs along three dimensions: diversity, relevance and language quality.

1. *Diversity*: refers to vocabulary richness, variety in expression and the extent to which the response is dissimilar from the rest in a generated collection of responses. distinct n-grams (Dist-n) [40], Entropy (Ent-n) [42] and Self-BLEU [43].
2. *Relevance* captures the extent to which the counterspeech addresses the central aspect of the hateful message and makes a coherent conversation towards mitigating the hate speech. For relevance, the authors compare (1) the generated response with the ground truth counterspeech by BLEU [44] and ROUGE [45] for syntactic similarity, and by MoverScore [46] and BERTScore [47] for semantic similarity; (2) the generated response with the hate speech by BM25, a relevance estimation function widely used in information retrieval.
3. *Language quality* quality measures whether the generated responses are grammatical, fluent and readable. The authors adopt GRUEN [48] to evaluate the language quality. Note that larger scores indicate better quality, except for Self-BLEU.

### 5.5.3 Results

Overall, GPS has the best diversity with significant margins than the baselines as noted in Table 5.3. For relevance, GPS has slightly better performance for BLEU, ROUGE, MoverScore and BERTScore, while has much better performance on BM25. This implies the counterspeech generated by GPS are more related to the hate speech and therefore, make more coherent conversations. The authors find that GPS is able to generate diverse and relevant rather than merely commonplace responses, such as "Please refrain from using such language". Therefore, the authors conclude that GPS has the best diversity and relevance, compared to the baselines. Besides, GPS has comparable language quality with the best baseline model—BART. Among these baselines, BART is the strongest one with much better relevance and language quality. Yet, BART still suffers from the diversity issue. Space-

Fusion has very poor results overall, though a manual inspection of the latent space fusion visualization suggests otherwise. One explanation is that SpaceFusion, with substantially more parameters compared with the Seq2Seq model may not have had sufficient training instances for its optimal performance. In their own experiments, the authors [37] demonstrate that SpaceFusion worked well on two datasets with 0.2M and 7.3M conversations, which is at least one to two orders of magnitude larger than their dataset. If provided with more training data, SpaceFusion could possibly be a strong candidate too. In comparison, though BART is an even more complicated model with 139M parameters, it was pre-trained on the BooksCorpus dataset with over 7,000 unique unpublished books and has the finetunable property.

## 5.6 CounterGeDi: A Controllable Approach to Generate Counterspeech

In most of the recent modules of counterspeech generation including the last one, the authors cannot control the generated output for any attribute from the vanilla generation model. However, as pointed out by different authors [49], counterspeech can vary based on the hate speech instance, demography of the hate and counter speakers [10] etc. Hence, the generation models without any control might produce suggestions that are not suitable for a particular instance. In this section, the authors try to understand if the authors can control the counter speech generation module and make the counterspeech more personalised.

### 5.6.1 Proposed Model

The basic module behind the proposed model is the GEDI model which is an unit which controls the generation of a larger model.

**GEDI**: For controlling generated counterspeech, the authors use a recent method Generative Discriminators (GEDI) [50], where the authors present a decoding time algorithm to control the output from the generation model. GEDI assumes the authors have class conditioned language model (CC-LM) with a desired control code $c$ and an undesired control code $\bar{c}$. For their case, the authors fix the control code $c$ as 'true' and $\bar{c}$ as 'false'. For each dataset, the attribute mentioned in the $+ve$ column in Table 5.4 is considered as desired, while $-ve$ column is used as undesired attribute.

The authors use the contrast between $P_\theta(x_{1:t}|c)$ and $P_\theta(x_{1:t}|\bar{c})$ to guide sampling from an LM that gives $P_{LM}(x_{1:T})$. The probability that the next token $x_t$ belongs to desired class is calculated using this contrast. This is calculated using Eq. 5.4 where $\alpha$ is a learnable scale parameter.

**Table 5.3** Automatic evaluation results. An asterisk * by the metric name indicates that the metric favors smaller values. Best results are in bold. LQ.: Language Quality; SB1: Self-BLEU-1; SB2: Self-BLEU-2; B2: BLEU-2; R2: ROUGE-2; MS: MoverScore; BS: BERTScore; GR: GRUEN [34]

| | | Diversity | | | | | | Relevance | | | | | LQ. |
|---|---|---|---|---|---|---|---|---|---|---|---|---|---|
| | | Dist-1 | Dist-2 | Ent-1 | Ent-2 | SB1* | SB2* | B2 | R2 | MS | BS | BM25 | GR |
| CONAN | Seq2Seq | 0.06 | 0.23 | 5.12 | 6.63 | 0.54 | 0.30 | 3.4 | 3.0 | 4.4 | 0.83 | 2.66 | 0.38 |
| | MMI | 0.06 | 0.23 | 4.88 | 6.41 | 0.57 | 0.35 | 2.9 | 2.3 | 3.9 | 0.82 | 1.63 | 0.33 |
| | SpaceFusion | 0.00 | 0.00 | 1.06 | 1.86 | 0.98 | 0.98 | 0.0 | 0.0 | −14.2 | 0.76 | 0.12 | 0.38 |
| | BART | 0.04 | 0.23 | 5.98 | 7.80 | 0.52 | 0.26 | 3.9 | 3.6 | 7.1 | 0.84 | 1.86 | 0.71 |
| | GPS | 0.06 | 0.27 | 5.77 | 7.41 | 0.43 | 0.19 | 7.1 | 6.5 | 10.9 | 0.85 | 5.43 | 0.71 |
| Reddit | Seq2Seq | 0.04 | 0.24 | 5.07 | 6.61 | 0.58 | 0.31 | 6.5 | 4.0 | 6.8 | 0.85 | 0.14 | 0.64 |
| | MMI | 0.05 | 0.32 | 5.11 | 6.76 | 0.56 | 0.29 | 6.4 | 4.0 | 6.9 | 0.85 | 0.14 | 0.56 |
| | SpaceFusion | 0.00 | 0.02 | 2.73 | 4.16 | 0.87 | 0.76 | 0.9 | 0.0 | −2.5 | 0.79 | 0.16 | 0.26 |
| | BART | 0.03 | 0.19 | 5.08 | 6.63 | 0.69 | 0.55 | 7.8 | 6.9 | 7.8 | 0.86 | 0.83 | 0.72 |
| | GPS | 0.09 | 0.53 | 5.74 | 7.61 | 0.41 | 0.15 | 8.1 | 7.1 | 7.8 | 0.87 | 2.58 | 0.75 |
| Gab | Seq2Seq | 0.02 | 0.17 | 5.14 | 6.71 | 0.56 | 0.30 | 7.5 | 5.0 | 6.7 | 0.86 | 0.14 | 0.67 |
| | MMI | 0.02 | 0.17 | 5.28 | 6.82 | 0.55 | 0.30 | 5.8 | 3.6 | 6.2 | 0.85 | 0.18 | 0.65 |
| | SpaceFusion | 0.00 | 0.01 | 3.72 | 4.84 | 0.81 | 0.73 | 1.8 | 0.1 | 0.0 | 0.82 | 0.17 | 0.21 |
| | BART | 0.03 | 0.17 | 5.42 | 7.25 | 0.60 | 0.38 | 6.9 | 6.4 | 6.8 | 0.86 | 0.81 | 0.72 |
| | GPS | 0.06 | 0.40 | 5.82 | 7.83 | 0.39 | 0.15 | 7.6 | 6.4 | 6.8 | 0.87 | 1.94 | 0.76 |

$$P_\theta (c \mid x_{1:t}) = \frac{P(c) P_\theta (x_{1:t} \mid c)^{\alpha/t}}{\sum_{c' \in \{c, \bar{c}\}} P(c') P_\theta (x_{1:t} \mid c')^{\alpha/t}} \quad (5.4)$$

In order to train the GEDI model, the authors combine the generative language modeling loss $\mathcal{L}_g$ (ref Eq. 5.5) as shown in Eq. 5.6 with a discriminative loss $\mathcal{L}_d$. The Eq. 5.7 shows the final loss $\mathcal{L}_{gd}$ where $\lambda$ is a learnable parameter.

$$\mathcal{L}_g = -\frac{1}{N} \sum_{i=1}^{N} \frac{1}{T_i} \sum_{t=1}^{T_i} \log P_\theta \left( x_t^{(i)} \mid x_{<t}^{(i)}, c^{(i)} \right) \quad (5.5)$$

$$\mathcal{L}_d = -\frac{1}{N} \sum_{i=1}^{N} \log P_\theta \left( c^{(i)} \mid x_{1:T_i}^{(i)} \right) \quad (5.6)$$

$$\mathcal{L}_{gd} = \lambda \mathcal{L}_g + (1 - \lambda) \mathcal{L}_d \quad (5.7)$$

For controlled generation, the authors propose a simple method to guide the model toward the target class which is represented using the heuristic equation 5.8 where $\omega$ is controllable parameter. In order to control multiple attributes, the authors extend the heuristic as represented in Eq. 5.9 where $\omega_i$ is a controllable parameter to bias the generation toward class $c_i$ for GEDI trained on the $i^{th}$ attribute.

$$P_w (x_t \mid x_{<t}, c) \propto P_{LM} (x_t \mid x_{<t}) P_\theta (c \mid x_t, x_{<t})^\omega \quad (5.8)$$

$$P_w (x_t \mid x_{<t}, c_1, ..c_n) \propto P_{LM} (x_t \mid x_{<t}) \prod_{i=1,...,n} P_\theta (c_i \mid x_t, x_{<t})^{\omega_i} \quad (5.9)$$

**Pipeline**

Their final operational setup involves three distinct components depicted in Fig. 5.4.

*Part A* signifies the base counterspeech generation model, initially trained on one of three counterspeech datasets. This section functions akin to an auto-regressive system, taking hate speech and the currently generated counterspeech (initially empty) to generate probability distributions for the subsequent token production.

*Part B* encompasses one or multiple GeDi models, each overseeing one of six total attributes. These GeDi models take the currently generated counterspeech as input and generate token probabilities specifically aligned with the desired attribute control. Initially, the authors allow the counterspeech generation models to generate 10 tokens without any constraints to provide the initial prompt to the GeDi model.

*Part C* selects the next token based on the token probabilities derived from various models—the counterspeech generation model and the GeDi models—following Eq. 5.9. In controlling each GeDi model, their primary manipulation is on the weight ($\omega$) as specified in Eq. 5.8, maintaining other parameters consistent with the methodology outlined in the paper [50].

## 5.6 CounterGeDi: A Controllable Approach to Generate Counterspeech

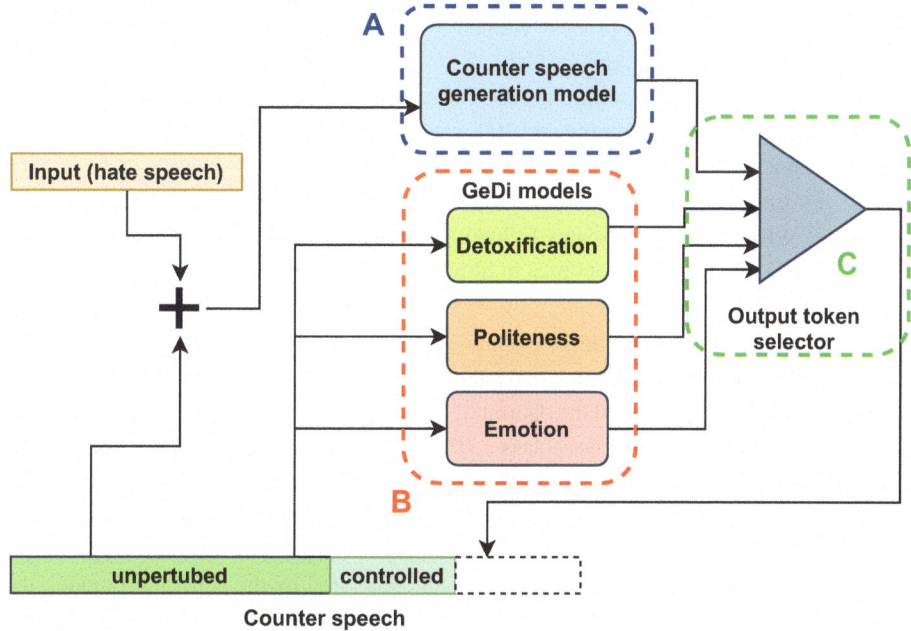

**Fig. 5.4** Overview of CounterGEDI model [51]

For a single attribute, the authors set the weight to 1, granting equal significance to both the counterspeech generation and the controlled attributes. For dual attribute control, weights are set at 0.5 for each attribute. In the case of triple attribute control, involving detoxification, politeness, and an emotional aspect, the authors assign weights of 0.3 to both politeness and detoxification and 0.4 to the emotional attribute. Additionally, the authors adopt nucleus sampling as their decoding strategy [33].

### 5.6.2 Experiments

**Datasets:** To assess their method, the authors utilized three public datasets containing hate speech and corresponding counterspeech instances. Specifically, the Reddit dataset comprises 5,257 instances, the Gab dataset has 14,614 instances [9], and the English portion of the CONAN dataset contains 408 instances [38].

In the Reddit and Gab datasets, counterspeech was authored by Amazon Mechanical Turk (AMT) workers, whereas the CONAN dataset enlisted expert NGO operators to provide counterspeech. The authors established pairs of hate speech and counterspeech from these datasets, ensuring that each hate speech instance was matched with an appropriate counterspeech response. This process yielded a total of 3,864 pairs for CONAN, 14,223 pairs for Reddit, and 41,580 pairs for Gab.

Consequently, the authors partitioned each dataset randomly into training, validation, and test sets, allocating 80% for training and 10% each for validation and testing, maintaining a consistent framework for robust evaluation.

*Attribute datasets*: the authors control several attributes in the generated counterspeech. The authors selected these attributes following the recommended strategies for counterspeech [18] and properties of responses in human conversation [52].

1. *Politeness*: One of the properties of counterspeech as suggested by the work [18] is empathy. As a first step in that direction the authors tried to make the generated counterspeech more polite. The authors used the dataset of 1.39 million posts released by Madaan et al. [53] labelled into nine politeness classes (P1-P9). As recommended by the authors, the authors considered P9 as the polite part and others (P1-P8) as non-polite.

2. *Detoxification*: This work [18] also noted several strategies that are discouraged while writing counterspeech. One of these discouraged strategies are hostile or aggressive behaviour. To detox any hostile counterspeech generated by the generation model, the authors use the a popular Kaggle dataset[5] which contains text samples having 'toxic' and 'non-toxic' labels. The authors stratified-split the released training dataset randomly into 90% training and 10% validation sets. The test set is already released separately with the dataset. The authors trained a GEDI model considering toxic as the positive label and non-toxic as the negative label. While generating using GEDI, the authors guide the generation toward the negative class (non-toxic).

3. *Emotion*: Another important aspect of conversation is communicating different emotions. A study [54] found that systems expressing emotions are more capable of providing user satisfaction. In the case of counterspeech, emotions might enhance the effect of the generated counterspeech [18]. For example, 'sadness' as an emotion can be added when the counter speakers affiliate themselves with the target group. Similarly, 'joy' can used to convey positivity in the counterspeech. In order to control the emotion while generating a counterspeech, the authors used a large dataset [55] of 416,809 datapoints comprising posts having seven emotions—'sadness', 'joy', 'fear', 'anger', 'surprise', and 'love'. For this paper, the authors did not consider—'love' and 'surprise' emotions as these had less than 10% posts in the dataset. The authors stratified-split each dataset randomly into training, validation, and test sets with 80% for training, and 10% for both validation and testing. The authors consider each emotion as a separate attribute and trained a GEDI model for that emotion by considering it as a positive label and other emotions as negative labels. For their experiments, the authors primarily focus on guiding the models toward the positive class.

---

[5] https://www.kaggle.com/c/jigsaw-toxic-comment-classification-challenge/data.

## 5.6 CounterGeDi: A Controllable Approach to Generate Counterspeech

**Table 5.4** This table shows the attribute datasets, positive and negative classes and data present in the train, validation and test part for each. $T_r$: Train, V: Validation, $T_e$: Test, p: polite, n-p: non-polite, t: toxic, n-t: non-toxic, s: sadness, j: joy, a: anger, f: fear, o: others. The % associated with the $T_r$, V and $T_e$ are the % of positive labels [51]

| Dataset | +ve | -ve | $T_r$ (%+ve) | V (%+ve) | $T_e$ (%+ve) |
|---|---|---|---|---|---|
| Polite | p | n-p | 1.12M (20%) | 137k (20%) | 137k (20%) |
| Toxic | t | n-t | 143k (10%) | 16k (10%) | 153k (4%) |
| Emotion | j | o | 333k (34%) | 42k (34%) | 42k (34%) |
|  | f | o | 333k (11%) | 42k (11%) | 42k (11%) |
|  | s | o | 333k (29%) | 42k (29%) | 42k (29%) |
|  | a | o | 333k (14%) | 42k (14%) | 42k (14%) |

A summary statistic of the attribute dataset for each of the tasks considered is noted in Table 5.4.

**Evaluation** The authors consider several metrics to evaluate their whole pipeline of controlled counterspeech generation. The *generation metrics* measure the generation capability of the DialoGPTm and GEDI models. The *classification metrics* are mainly to evaluate the GEDI model on the attribute datasets. Finally, the authors measure the amount of control in the generated counterspeech using external classifiers which the authors refer to as *controller metrics*. The authors generate 5 samples for every hate speech instance with DialoGPTm. The GPS framework automatically selects the best response based on the heuristic, hence the authors keep one sample for every hate speech instance.

1. **Generation metrics**: To measure the generation quality, the authors use different standard metrics. The authors use *BLEU-2*[6] and *METEOR* [56] to measure how similar the generated counterspeech are to the ground truth counterspeech. The authors also measure if the generation model generates a diverse and novel counterspeech using metrics from previous research [17]. To measure fluency, the authors use a classifier of linguistic acceptability trained on the COLA dataset [57].
2. **GEDI metrics**: For classification, the authors report *accuracy*, *macro F1-score*, and *AUROC* score for each GEDI model's performance on a test dataset of a particular attribute. The authors also report the generation performance using the perplexity [27].
3. **Controller metrics**: In order to evaluate the ability of the GEDI controller to control the attribute, the authors used third-party classifiers for each attribute. For politeness, the authors trained a bert-base-uncased model for politeness level detection on a scale of 0–7.[7] For measuring emotion in the generated text, the authors used the Ekman version

---
[6] Converted to a scale of 0–100 from 0–1.
[7] https://github.com/AlafateABULIMITI/politeness-detection.

**Table 5.5** Evaluation results for the three datasets. The authors report BLEU-2 (B2), COLA, METEOR (M), novelty (N) and diversity (D) to compare the two baselines: generate-prune-select (GPS) framework and DialoGPTm. For all metrics, higher is better and **bold** denotes the best scores [51]

| Model | B2 (↑) | COLA (↑) | M (↑) | N (↑) | D (↑) |
|---|---|---|---|---|---|
| *CONAN* | | | | | |
| GPS | **41.5** | **0.82** | 0.14 | 0.18 | 0.60 |
| DialoGPTm | 12.7 | 0.78 | **0.18** | **0.84** | **0.80** |
| *Reddit* | | | | | |
| GPS | **14.1** | **0.82** | 0.11 | 0.30 | 0.47 |
| DialoGPTm | 6.9 | 0.75 | **0.17** | **0.82** | **0.74** |
| *Gab* | | | | | |
| GPS | **13.9** | **0.82** | 0.12 | 0.15 | 0.41 |
| DialoGPTm | 7.7 | 0.80 | **0.17** | **0.80** | **0.72** |

of the GoEmotions models.[8] For each post, it returns a confidence score between 0 and 1 for anger, disgust, fear, joy, sadness, surprise + neutral. the authors report the confidence score for a particular emotion as a measure of that emotion in a given post. Finally, to measure toxicity the authors used the HateXplain model [20] trained on two classes—toxic and non-toxic.[9] The authors report the confidence between 0 and 1 for the non-toxic class.

### 5.6.3 Results

*Generation results*: the authors compared the DialoGPTm model with GPS in Table 5.5. While GPS shows better **BLEU-2** scores across all three datasets, DialoGPTm outperforms with higher **METEOR** scores. Additionally, DialoGPTm demonstrates greater novelty and diversity for these datasets. In terms of fluency measured by the **COLA** metric, GPS excels as it filters out grammatically incorrect samples. Considering DialoGPTm's competitive performance, akin to the state-of-the-art model, the authors opt to proceed with DialoGPTm for further experiments.

**GEDI metrics**: As reported in Table 5.6, the authors find that F1-score and AUCROC scores for politeness and all the four emotions are above 0.9. This highlights that even with 0.2 as the weight for the discriminator the authors are able to get good scores on classification.

---

[8] https://huggingface.co/monologg/bert-base-cased-goemotions-ekman.
[9] https://huggingface.co/Hate-speech-CNERG/bert-base-uncased-hatexplain-rationale-two.

## 5.6 CounterGeDi: A Controllable Approach to Generate Counterspeech

**Table 5.6** GEDI generation and classification performance on test set of attribute datasets. Generation is evaluated using the Perplexity whereas classification performance is measured using F1-score (F1), Accuracy (Acc) and AUCROC (AUC). For all the metrics except perplexity, higher is better [51]

| Dataset | Positive | F1 (↑) | Acc (↑) | AUC(↑) | Perplexity (↓) |
|---|---|---|---|---|---|
| Toxicity | toxic | 0.60 | 0.85 | 0.84 | 4.428 |
| Politeness | polite | 0.93 | 0.96 | 0.93 | 3.476 |
| Emotion | joy | 0.96 | 0.96 | 0.97 | 3.546 |
| Emotion | sadness | 0.98 | 0.98 | 0.99 | 3.543 |
| Emotion | fear | 0.94 | 0.97 | 0.98 | 3.774 |
| Emotion | anger | 0.96 | 0.98 | 0.99 | 3.560 |

The perplexity scores for all the test datasets are also around 3.5.[10] GEDI model for toxicity has lower scores than the other attribute tasks. The F1-score for toxicity detection is ∼ 0.6 and AUCROC is ∼0.83. The perplexity is also higher at around 4.5 for the toxicity dataset. This highlights the difficulty of the task of detecting toxicity.

*Single-attribute control*: In Table 5.7, the authors report the amount of different attributes present in the generated counterspeech for each dataset and for each model. When the authors compare GPS and DialoGPTm, the authors find that except for anger emotion, all other scores are significantly higher for DialoGPTm. Second, using a control for a particular attribute significantly improves the presence of that attribute (p-value <0.001). For instance, in Table 5.7, the politeness score increases from 3.91 to 4.54, from 5.24 to 6.05 and 5.14 to 6.11 for CONAN, Reddit and Gab respectively when the DialoGPTm model is controlled for politeness. This is true for all attributes barring the 'anger' emotion. Politeness and detoxification scores increased by 15–18% and 6–8% respectively across all the datasets. For the emotion attributes, 'joy' has the highest scores among all for both controlled and uncontrolled attribute. The authors see an overall increase in 'joy' of around 17% for Gab, 14% for Reddit and 88% for CONAN. Counter responses in CONAN datasets are mostly devoid of any emotions hence bringing a change in them is much easier than the Reddit/Gab datasets which are higher in terms of the joy attribute. The authors reach closer to GPS baseline for anger emotion while controlling anger emotion and increase the score by 54%, 55% and 16% for Reddit, Gab and CONAN, respectively. While the increase for other emotions—'sadness' and 'fear' increased significantly, the overall scores for them remain low.

*Multi-attribute control*: the authors also generate counterspeech with the DialoGPTm with mutli-attribute control. The authors keep politeness, detoxification and one of the emotion[11] as control attributes. This gives us four variations for each dataset. The authors then measure the individual attribute scores for each of these three attributes and report the results in

---

[10] For reference, perplexity for pretraining GPT-2 comes around 10 after 10K steps (https://tinyurl.com/3vwrvscd).

[11] One among 'joy', 'anger', 'fear' and 'sad'.

**Table 5.7** Performance of single attribute setups with the vanilla baseline generate-prune-select (GPS) and DialoGPTm models. Each column name represents the attribute being measured. The attributes measured are politeness (P), detoxification (D), sadness (S), joy (J), anger (A) and fear (F). Politeness (P) is measured on a scale of 0–7 whereas others are measured on the scale [0, 1]. For the last row—controlled DialoGPTm (DialoGPTm-c) the column name also represents the attribute getting controlled. For all the metrics, higher is better and **bold** denotes the best scores [51]

| Model | D (↑) | P (↑) | J (↑) | A (↑) | S (↑) | F (↑) |
|---|---|---|---|---|---|---|
| *CONAN* | | | | | | |
| GPS | **0.68** | 2.01 | 0.16 | **0.12** | 0.03 | 0.01 |
| DialoGPTm | 0.64 | 3.91 | 0.18 | 0.09 | 0.04 | 0.01 |
| DialoGPTm-c | **0.68** | 4.54 | 0.34 | 0.11 | **0.08** | **0.05** |
| *Reddit* | | | | | | |
| GPS | 0.82 | 1.62 | 0.23 | **0.32** | 0.04 | 0.01 |
| DialoGPTm | 0.82 | 5.24 | 0.63 | 0.17 | 0.06 | 0.00 |
| DialoGPTm-c | **0.87** | **6.05** | **0.72** | 0.27 | **0.10** | **0.02** |
| *Gab* | | | | | | |
| GPS | 0.79 | 1.46 | 0.22 | **0.28** | 0.04 | 0.01 |
| DialoGPTm | 0.81 | 5.14 | 0.66 | 0.17 | 0.05 | 0.00 |
| DialoGPTm-c | **0.85** | **6.11** | **0.77** | 0.26 | **0.10** | **0.02** |

Table 5.9. For detoxification scores, the setup—$joy + polite + detox$ outperforms other setups across all the experiments. This setup even outperforms the single-attribute detoxification setup by 8%, 2% and 2% for CONAN, Reddit and Gab, respectively. For the politeness score, the best performance occurs for $joy + polite + detox$ setup for CONAN and Reddit dataset, while the setup—$fear + polite + detox$ performs better in case of the Gab dataset. Compared to the single-attribute setup for politeness, the politeness scores drop across all the multi-attribute setups. Among the emotions, the attribute score for 'joy' in a multi-attribute setting outperforms the single attribute setting by 44%, 13% and 10% for CONAN, Reddit and Gab. For 'anger', the scores in the multi-attribute setting decrease around 25–30% when compared to the single attribute setting. For other attributes like 'sadness' and 'fear', the multi-attribute results are below 0.1, similar to the single attribute results. Please also see the Appendix for attribute ablation performances.

*Quality of controlled generation*: In the previous section, the authors observed that the authors were able to control attributes in generated outputs in single and multi-attribute setups. While this is encouraging, it is important to understand if the controlled text are losing the central theme of remaining a counterspeech and are still fluent. For the former, the authors measure the **BLEU-2** metric and for the latter the authors use the **COLA** metric.

According to Table 5.8, the authors find that the relevance of the output (measured using BLEU-2) does not change much across different attributes for the single attribute setups. For

**Table 5.8** BLEU-2 and COLA performance for single attribute setups for DialoGPTm-c model. Each column name represents the individual attribute model namely politeness (P), detoxification (D), sadness (S), joy (J), anger (A) and fear (F). **Bold** denotes the best scores across the row

| Scores | Detox | Polite | Joy | Anger | Sadness | Fear |
|---|---|---|---|---|---|---|
| CONAN | | | | | | |
| BLEU-2 | **13.8** | 12.1 | 12.2 | 11.6 | 12.0 | 12.8 |
| COLA | **0.83** | 0.72 | 0.72 | 0.74 | 0.76 | **0.72** |
| Reddit | | | | | | |
| BLEU-2 | **8.1** | 7.8 | 7.7 | 7.8 | 7.5 | 7.3 |
| COLA | 0.72 | 0.77 | 0.70 | 0.72 | **0.81** | 0.70 |
| Gab | | | | | | |
| BLEU-2 | **8.7** | 8.3 | 8.5 | 8.3 | 8.2 | 8.3 |
| COLA | **0.85** | 0.82 | 0.76 | 0.76 | 0.80 | 0.78 |

some of the attributes like detoxification, the BLEU-2 scores even outperform the DialoGPTm model (without control) for all the datasets as noted in column B2 in Table 5.5. For Reddit and Gab, there is a further improvement of 1–2 points in the **BLEU-2** metric for other attributes also as compared to the vanilla DialogGPTm model (in column B2 in Table 5.5). This shows that the controls do not affect the overall relevance of the generated counterspeech. In fact, the relevance improves in many cases. In terms of fluency, the authors see a slight drop which comes as a cost for controlling different attributes except few cases (comparing column COLA in Tables 5.5 and 5.8). This might be due to the fact that GEDI model is not geared toward maintaining the fluency of the models. The observation holds for the multi-attribute setup as well (comparing columns B2 and COLA in Tables 5.5 and 5.9).

Overall, the authors observe that it is possible to control the attributes in the generated outputs using the single attributes. Their experiments with multi-attributes further reveal that there are certain complementing attributes e.g., $joy + polite + detox$ which can be used to further increase the single-attributes setups. For other setups, the attribute scores drop below the single attribute setups. Another promising observation is that the control of attributes does not harm the relevance of the generated output as they still remain close to the ground truth. Since GEDI is not geared toward improving fluency, the authors see a slight drop in the fluency of the generated outputs. An interesting research direction would be to look into improving attribute and fluency scores while using multi-attribute setups.

## 5.7 Knowledge-Grounded Counter Speech Generation

Many counterspeech contain one or more factual arguments in order to counter the hate content. But the facts generated by the model might be wrong which results in wrong information. This is commonly known as **hallucination**. The current definition of hallucinations

**Table 5.9** Results of controlling three attributes—politeness, detoxification and one of the emotions in a multi-attribute setting. The columns represent the amount of the attribute present for each setup. The column—*emotion* represents the score of the emotion shown in the parenthesis that is being controlled for that instance. BLEU(B2) and COLA were also reported for different setups. For all metrics, higher is better and **bold** denotes the best scores

| Attributes | Detox(↑) | Polite(↑) | Emotion(↑) | B2(↑) | COLA(↑) |
| --- | --- | --- | --- | --- | --- |
| *CONAN* | | | | | |
| Joy(J)+P+D | **0.74** | **4.13** | 0.49 (J) | 13.4 | **0.79** |
| Anger(A)+P+D | 0.67 | 3.06 | 0.08 (A) | 12.6 | 0.68 |
| Sad(S)+P+D | 0.70 | 3.56 | 0.07 (S) | 13.2 | 0.74 |
| Fear(F)+P+D | 0.70 | 4.00 | 0.06 (F) | **13.6** | 0.75 |
| *Reddit* | | | | | |
| Joy+P+D | **0.89** | **5.79** | 0.82 (J) | 8.3 | **0.81** |
| Anger+P+D | 0.85 | 4.24 | 0.19 (A) | **8.3** | 0.72 |
| Sad+P+D | 0.87 | 3.56 | 0.09 (S) | 8.2 | 0.79 |
| Fear+P+D | 0.87 | 4.00 | 0.01 (F) | 7.8 | 0.79 |
| *Gab* | | | | | |
| Joy+P+D | **0.87** | 5.68 | 0.85 (J) | **8.8** | **0.85** |
| Anger+P+D | 0.83 | 4.11 | 0.19 (A) | 8.5 | 0.75 |
| Sad+P+D | 0.85 | 4.70 | 0.09 (S) | **8.8** | 0.84 |
| Fear+P+D | 0.86 | **5.82** | 0.01 (F) | **8.8** | 0.83 |

**Table 5.10** Hallucinated CN generated by GPT-2 that is fluent and credible (according to Office for National Statistics, the Muslim population is just above 3M)

| HS: Muslims are invading our country |
| --- |
| CN: Actually, there were 16,938,000 Muslims in the UK in 2016, so if you exclude London, that is actually increasing by 2%, which doesn't seem very significant |

aligns with prior research [8], characterizing them as generated content that is nonsensical or unfaithful to the provided source content. A hallucinated example of a fact-bound counterspeech, generated by GPT-2 model fine-tuned as done by the work [58], is provided in Table 5.10. In this section, the authors try to develop a model that can use external knowledge to improve the counterspeech model and do less hallucination [59].

## 5.7.1 Proposed Model

Their depicted architecture in Fig. 5.5 comprises a knowledge retrieval segment responsible for retrieving relevant knowledge at the sentence level, along with a generation module for creating counterspeech. Initially, the knowledge retrieval module prepares different versions of a query Q for a given hate speech HS using two strategies: query extraction ($Q_{hs}$) and automatic query generation ($Q_{gen}$). Subsequently, these obtained queries are utilized to search for pertinent knowledge articles through a search engine. Finally, a sentence selector is employed to filter and rank the most pertinent sentences from the retrieved articles, forming the relevant knowledge (KN).

Regarding the counter-narrative generation module, the authors fine-tuned multiple language models. These models take both hate speech and the ranked knowledge sentences KN as inputs and generate corresponding counter-narratives as outputs.

#### 5.7.1.1 Knowledge Retrieval Module

The knowledge retrieval module in the architecture incorporates a knowledge repository, a query construction sub-module, and a knowledge sentence selection sub-module.

*Knowledge repository*: Previous methods for incorporating external knowledge into dialogue generation have utilized both structured and unstructured knowledge. However, due to the absence of structured knowledge in the hate speech domain, the authors rely on unstructured textual knowledge, specifically articles, which allows for easy updates to the knowledge repository. Since hate speech often emerges in response to target-related events like terrorist attacks, the ability to update knowledge, such as news articles, enables us to create appropriate counter-narratives containing the latest statistics or evidence from current events. The authors augment their knowledge repository with Newsroom [60] and WikiText-103 [61]. WikiText-103 encompasses a vast collection of 28,595 complete Wikipedia articles spanning over 103 million words. Meanwhile, Newsroom comprises 1.3 million articles extracted from major news publications between 1998 and 2017, totalling over 6.9 million words.

*Query construction*: To construct comprehensive and proper queries to search for relevant knowledge for the data pairs, the authors applied two strategies: (i) **query extraction** and (ii) **query generation**. In both strategies, the query is composed of keyphrases that can be defined as important and topical phrases from a text [62].

- *Query extraction*: the authors extracted keyphrases from CONAN dataset using Keyphrase Digger [63], a multilingual keyphrase extraction systems that uses statistical measures and linguistic information, and is proven to be one of the best systems for unsupervised settings3. Following the knowledge retrieval strategy using input argument [64] for counterargument generation, the authors first obtained the hate speech key phrases to construct the initial query $Q_{hs}$. However, hate speech from CONAN mostly contains hateful and simplistic phrases in comparison to the input arguments used [64] that can be rich in con-

**Fig. 5.5** Architecture of knowledge grounded generation with extracted (green solid arrow) and generative (dotted arrow) queries (topical phrases) that are exploited to retrieve relevant knowledge. The knowledge sentences extracted together with input HS are fed to CN generation. The authors give an example of a generative approach

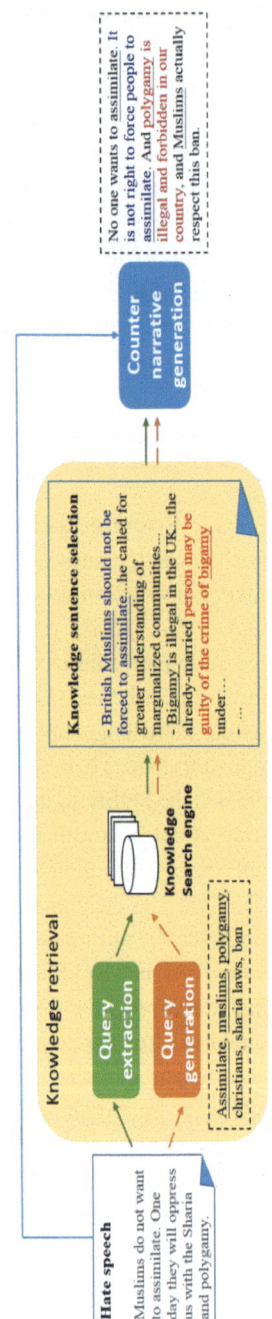

## 5.7 Knowledge-Grounded Counter Speech Generation

tent. Therefore, in the hate speech-counterspeech scenario, the authors hypothesize that the keyphrases from $Q_{hs}$ alone would not be sufficient for relevant knowledge search, especially for mapping the knowledge onto training data. To this end, the authors also extracted keyphrases from counterspeech together with hate speech to increase the possibility that the retrieved knowledge sentences contain pieces of information found in the ground truth. Hence, the second query $Q_{hsUcn}$ contains counterspeech keyphrases for the relevancy to the target counterspeech and hate speech key phrases for preserving the hate context.

- *Query generation*: Since the best query configuration $Q_{hsUcn}$ cannot be available at test time, the authors need a way to obtain keyphrases that serve as counterspeech cues for searching knowledge sentences during the counterspeech generation. To this end, the authors built a query generation model that takes HS as input and outputs a comma-separated list of counterspeech keyphrases, which is then used as $Q_{gen}$. They aim to obtain an approximation of $Q_{hsUcn}$ via $Q_{hsUgen}$ at the test time. The model is trained using Transformer [65] architecture and the authors use the CONAN dataset [38].

*Knowledge sentence selection* the authors use Solr5 to index the articles and retrieve those relevant to a given query based on the similarity between the articles and the query using BM25 [66]. Once the queries have been obtained either through extraction or generation, they are presented to Solr for retrieving the 25 top-ranked articles. Next, the authors used spaCy sentence segmentation6 to split an article into sentences. Similar to this work [67], given a query Q the authors score each sentence xi in the set of articles D independently, using ROUGEL-F1.

In the final step, the authors distilled the knowledge by keeping the top 5 knowledge sentences that have the highest scores among the 25 top-ranked articles. Instead of more stringent filtering, such a setting has been applied to grant a better variety of source articles and corresponding distilled sentences. The authors refer to such automatically associated sentences as "silver knowledge".

### 5.7.2 Experiments

Large pretrained LMs require less amount of high quality data to be fine-tuned on downstream tasks while providing strong performances and they already store a large amount of factual and commonsense knowledge from their training data [68]. To this respect, the authors built the following models:

- $GPT2_{KN}$—obtained by fine-tuning GPT-2 on CONAN data paired with KN.
- $GPT2_{KN,MT}$—by fine-tuning GPT-2KN in a multi-task learning fashion for learning to distinguish counterspeech from hate speech as next utterances.

- *XNLG* [69]—This is a pre-trained Transformer-based language model trained on Wikipedia dumps with two relevant objectives for their task: to obtain contextual representations and to recover a given input.

The authors expect all three models to attend over the hate speech and retrieve knowledge and look for the relevant snippets to be recovered while generating a counterspeech. The training HS-CN pairs are represented as HS [HS end token] KN [KN end token][12] CN [CN end token]. Each model is trained with $Q_{hsUcn}$ and then tested on $Q_{hs}$, $Q_{gen}$, and $Q_{hsUgen}$. The authors use the following baselines for this comparison.

- Non pretrained *Transformer* without knowledge using the same hyper-parameters as key phrase generation model.
- *GPT-2* without knowledge following the same configuration as GPT-2KN.
- *Candela* [64] an LSTM-based state-of-the-art knowledge-driven architecture for argument generation. The authors hypothesize that a pre-training procedure9 on data from a similar task (argument generation) can be beneficial for generalization and porting knowledge.

### 5.7.3 Results

the authors report BLEU-2 (B-2) and ROUGE-L (R-L) scores for all proposed models and baselines in Table 5.11 on the test split of CONAN that the authors automatically paired with silver knowledge using various queries. The authors also measure the capability of each model to produce *novel* responses with respect to the training data by Jaccard similarity [17], and *diverse* responses for the given input by repetition rate (RR) [70].

Among their models, GPT-$2_{KN}$ achieves the highest B-2 score, while XNLG stands out for its exceptional novelty, diversity, and R-L score. The significant improvement in novelty seen in knowledge-grounded models underscores the advantages of incorporating knowledge for generating CNs compared to the baseline models, especially Transformer TRF.

However, despite XNLG's strong quantitative performance, a closer examination of its output reveals that it tends to mimic almost everything from the knowledge-driven model (KN) rather than generating counterspeech authentically, thereby inflating novelty and diversity scores. This issue becomes evident when comparing the average word and sentence counts in the outputs of XNLG with those of other models detailed in Table 5.11.

---

[12] https://spacy.io/universe/project/spacy-sentence-segmenter.

## 5.8 LLMs for Mitigation

**Table 5.11** Results of CN generation with silver knowledge. The authors report novelty (Nov.), RR, BLEU-2 (B-2), ROUGE-L (R-L), KN overlap with generation and the average amount of words and sentences per generation

| Models | Nov. | RR | B-2 | R-L | #Word | #Sent. | KN overlap (n-gram) | | |
|---|---|---|---|---|---|---|---|---|---|
| | | | | | | | 1 | 2 | 3 |
| *Without knowledge* | | | | | | | | | |
| TRF | 0.467 | 7.72 | 0.082 | 0.094 | 21.47 | 1.70 | – | – | – |
| GPT-2 | 0.688 | 9.04 | 0.045 | 0.100 | 15.95 | 1.35 | – | – | – |
| Train$_{cn}$ | – | 3.91 | – | – | 21.79 | 1.87 | 0.307 | 0.054 | 0.016 |
| *With knowledge* | | | | | | | | | |
| Candela ($Q_{hs}$) | 0.692 | 21.87 | 0.040 | 0.098 | 23.85 | 2.47 | 0.173 | 0.008 | 0.001 |
| **GPT-2$_{KN}$** | | | | | | | | | |
| w/ $Q_{hs}$ | 0.723 | 8.13 | 0.082 | 0.094 | 15.60 | 1.32 | 0.258 | 0.023 | 0.008 |
| w/ $Q_{gen}$ | 0.728 | 7.48 | 0.067 | 0.091 | 12.75 | 1.17 | 0.260 | 0.050 | 0.019 |
| w/ $Q_{hs \cup gen}$ | 0.735 | 6.30 | 0.085 | 0.103 | 15.35 | 1.59 | 0.358 | 0.068 | 0.024 |
| w/ $Q_{hs \cup cn}$ | 0.727 | 7.17 | **0.166** | 0.110 | 13.10 | 1.16 | 0.282 | 0.058 | 0.022 |
| **GPT-2$_{KN,MT}$** | | | | | | | | | |
| w/ $Q_{hs}$ | 0.744 | 11.69 | 0.050 | 0.090 | 13.35 | 1.17 | 0.269 | 0.049 | 0.017 |
| w/ $Q_{gen}$ | 0.731 | 10.37 | 0.052 | 0.092 | 13.34 | 1.14 | 0.253 | 0.044 | 0.017 |
| w/ $Q_{hs \cup gen}$ | 0.747 | 7.59 | 0.091 | 0.090 | 16.91 | 1.26 | 0.269 | 0.033 | 0.009 |
| w/ $Q_{hs \cup cn}$ | 0.731 | 9.56 | 0.048 | 0.107 | 13.05 | 1.13 | 0.276 | 0.057 | 0.023 |
| **XNLG** | | | | | | | | | |
| w/ $Q_{hs}$ | **0.824** | 14.42 | 0.073 | 0.084 | 55.51 | 3.71 | 0.841 | 0.650 | 0.558 |
| w/ $Q_{gen}$ | 0.819 | 6.88 | 0.097 | 0.084 | 55.64 | 3.64 | 0.849 | 0.656 | 0.558 |
| w/ $Q_{hs \cup gen}$ | 0.812 | 6.98 | 0.074 | 0.089 | 57.58 | 3.00 | 0.828 | 0.579 | 0.475 |
| w/ $Q_{hs \cup cn}$ | 0.819 | **5.69** | 0.076 | **0.116** | 55.69 | 3.42 | 0.840 | 0.631 | 0.529 |

GPT-2$_{KN,MT}$ lags behind their other models in RR, B-2, and R-L metrics but still maintains a competitive novelty score. As for Candela, while it demonstrates similar performance levels in terms of R-L and B-2, its generated content tends to be repetitive and lacks novelty.

## 5.8 LLMs for Mitigation

So far, the authors have explored how the mitigation systems existed before the eruption of large language models (LLMs). As mentioned in the detection part, LLMs can definitely play a role in improving the moderation pipeline, including the mitigation pipeline. One way in which LLMs can be used in the mitigation pipeline is to generate counterspeech.

In this book, we investigate the applicability of four LLMs (GPT-2, DialoGPT, FlanT5 and ChatGPT) for zero-shot counterspeech generation, which is the first ever attempt of this kind. We evaluate these models over four different counterspeech generation dataset—CONAN [38], CONAN-MT [22], Reddit and Gab [71]. We also try to develop different prompts to control the type of counterspeech. In the next section, we discuss the models we used for generating counterspeech in zero-shot setting.

### 5.8.1 Models Used

**GPT-2** [26] is trained on a large dataset called WebText. The dataset contains the textual content found in 45 million links shared by users on Reddit. Note that, WebText is not directly sourced from Reddit itself, but rather consists of data derived from outbound links posted on Reddit. This language model is trained with the objective of predicting the next word, given all the previous words within some text. The model aims at maximising $p(x) = \prod_{i=1}^{n} p(x_n|x, ..., x_{n-1})$, for our experiments we use all three different versions of GPT-2—117M (small), 345M (medium), and 762M (large) parameters from this link.[13]

**DialoGPT** [27] is trained on a large corpus consisting of English Reddit dialogues. The corpus consists of 147 million instances of dialogues, collected over 12 years. Unlike GPT-2, this model generates better dialogue-like responses to any given prompt. In this model, along with ground truth response $T = x_1, ..., x_n$, we also have a dialogue utterance history $S$. The model aims at maximising $p(T|S) = p(x_1|S) \prod_{i=2}^{n} p(x_i|S, ..., x_{i-1})$. For our experiment, we use all three different versions of DialoGPT—117M (small), 345M (medium), and 762M (large) parameters from this link.[14]

**FlanT5** [72] is a T5 [73] model that has been finetuned on a multi-task mixture of supervised tasks and for which each task is converted into a text-to-text format. They use instruction finetuning [74] procedure on each of these tasks, as well as chain-of-thought (CoT) prompting [75]. Overall, the authors show that using such a framework improves the results across various benchmarks over the T5 versions.

**ChatGPT** [76] is trained with a GPT 3.5 model using Reinforcement Learning from Human Feedback (RLHF), using a similar method as InstructGPT [74] but with slight differences in the data collection setup. ChatGPT performs exceptionally well in question-answering scenarios. Beyond its capability of being a conversational tool, many attempts have been made to evaluate the quality of ChatGPT-generated texts in various domains [77].

---

[13] https://huggingface.co/docs/transformers/model_doc/gpt2.
[14] https://github.com/microsoft/DialoGPT.

## 5.8.2 Prompting Strategies

### 5.8.2.1 Vanilla Prompting

In this scenario, we just ask the model to generate a counterspeech using a very simple prompt—"Write a counterspeech to the given hate speech". We did not add additional constraints because we wanted to measure the intrinsic properties of the models.

### 5.8.2.2 Type-Specific Prompting

Type specific generation is another challenge in counterspeech generation [10]. In this regard, the first few words of the counterspeech can be essential for the type we want to generate. We propose three different prompting strategies to generate the first few words (prompts). These prompts help us in controlling the type of the counterspeech generated by the above LLMs.

**Manual prompting**: In this strategy, two authors experienced in hate speech detection research read through the prompts dataset and created 2-3 possible beginnings (prompts) for each type of counterspeech. These prompts guide the model to generate the appropriate type of counterspeech. For manual prompts, we do not set a hard limit on the number of words but asked the authors to make them small.

**Frequency based prompting**: In this strategy, we collect the beginning four words as a sub-string from each counterspeech of each type and cluster using exact matching of the sub-strings. These sub-strings represent the prompts for a particular type. We take the top five prompts based on their frequency for each type.

**Cluster centered prompting**: In this strategy, we first pass all the counterspeech from the prompting dataset through sentence embedding model `all-mpnet-base-v2`.[15] For each type of counterspeech we then cluster the embeddings using $K$-means clustering. We decide the number of cluster for each type of counterspeech using the elbow method [78]. We note the number of clusters for each type in Table 5.13. The average clusters are ~16 per type. Out of these clusters we select the top 10 clusters using their cluster size. Then we select top three sentences per cluster which lie closest to the cluster center. These sentences act as a representative of that cluster. The final prompts per cluster are comprising the beginning 4 words as a sub-string from each of these three sentences. This way we collect 30 prompts per type in total.

We note one instance of prompts for each strategy per type in the Table 5.12.

---

[15] https://huggingface.co/sentence-transformers/all-mpnet-base-v2.

**Table 5.12** One instance of prompt using each prompt strategy per type. Each column represents the different prompt strategies and each row represents a particular type of counterspeech. M: Manual, CC: Cluster centered, FB: Frequency based. F: Facts, Hy: Hypocrisy, Hu: Humour, Af: Affiliation, Q: Questions, D: Denouncing

| Type | M | CC | FB |
|------|---|----|----|
| F  | This is a fact    | The myth that muslims | The vast majority of |
| Hy | In contradiction  | I am wondering have   | If you are really    |
| Hu | This is funny     | I bet she got         | Must be hard for     |
| A  | I also belong     | I am jewish and       | I am a christian     |
| Q  | Are you aware of  | How do you know       | Why do you think     |
| D  | Please do not say | Why is the hate       | Why is this a        |

### 5.8.3 Additional Datasets

In this section, we describe the additional datasets required for the evaluation of the generated responses and creating of prompts.

**Counterspeech dataset**: For measuring the quality of counterspeech, we use two datasets from past works [10, 79]. Both these datasets have type information, the paper [10] in addition have non-counterspeech comments curated from YouTube. We use two variations of the counterspeech dataset.

The first variant compiles the counterspeech posts themselves. This is primarily used for **classification** of a text into a counterspeech or not. In order to align our settings with the recommendations given by the paper [80], we place the hostile category counterspeech in the non-counterspeech part in this variant [10]. This way, we had 4,175 counterspeech comments and 9,765 non-counterspeech comments. We divide the dataset into train:validation:test in the ratio of 8:1:1 using stratified sampling.

The second variant compiles the counterspeech types. This is primarily used for **classification** of a counterspeech into one of its types and **prompting**. We take the counterspeech posts from the paper [10] and merge them with the counterspeech from the paper [79] along with their types. One thing to note is that we don't utilise the posts from the paper [10] which contain more than one strategies of counterspeech. Next, we study the definition of different types of counterspeech and select six types of counterspeech which appeared distinctive to all the authors unanimously. The statistics of the dataset finally extracted is noted in Table 5.13.

We divide the dataset, with 30% of it for constructing **prompts** and the rest for classification using stratified sampling. The classification dataset was further divided into train:validation:test in the ratio of 8:1:1 using stratified sampling. The prompt dataset is used to create the type prompts.

## 5.8 LLMs for Mitigation

**Table 5.13** This table represents the type-specific information for each type of counterspeech that we considered for our task. The columns *Prompting* and *Classification* represent the amount of data points used for finding prompt strategies and building classification model. *F1 score* (F1) column shows the performance of the type classifier. *Clusters* column represents the number of clusters found per type using the cluster-centered prompting strategy

| Type | Classification | Prompting | F1 | Clusters |
|---|---|---|---|---|
| Hypocrisy | 579 | 248 | 0.59 | 15 |
| Denouncing | 738 | 316 | 0.85 | 16 |
| Humor | 607 | 260 | 0.76 | 16 |
| Facts | 1094 | 469 | 0.84 | 18 |
| Affiliation | 163 | 70 | 0.84 | 16 |
| Question | 227 | 97 | 0.97 | 18 |
| Average/total | 3408 | 1460 | 0.80 | 16.5 |

**Counterargument dataset**: For evaluating the counterargument quality we select a popular argument dataset [81] which has 6,317 against and 4,822 for arguments categorized into six topics. For each topic, we assume all possible pairs of arguments. From this set, we sample (without replacement) 10,000 pairs which have the same stance and 10,000 pairs which have the opposite stance. This way, we have a dataset of 60,000 argument pairs. We divide the dataset into train:validation:test in the ratio of 8:1:1 using stratified splitting.

### 5.8.4 Evaluation Metrics

*Generation metrics*: To measure the generation quality, we use different standard metrics. We use `gleu` [82] and `meteor` [83] to measure how similar the generated counterspeech are to the ground truth references. We also measure if the LLMs generates diverse and novel counterspeech. For this purpose, we use the `diversity` and `novelty` metrics from existing literature [17]. In addition, we also report one of the recent generation metrics, `bleurt` [84]. Note that, we do not use the BLEU [85] score because it has some undesirable properties when used for single sentences, as it is designed to be a corpus-specific measure [82]. Further, the reader might notice negative scores in the case of `bleurt` metric. This is not unnatural since the `bleurt`, unlike BLEU, is not calibrated. For more information, refer here.[16]

*Engagement prediction metrics*: We use the DialogRPT model [19] to predict the human feedback of the counterspeech generated using the following metrics—*width*: the number of direct replies to the given reply, *depth*: the maximum length of dialogue after this turn, and *updown*: the number of upvotes minus the number of downvotes.

---

[16] https://github.com/google-research/bleurt/issues/1.

This metric can help us in identifying how engaging the generated counterspeech is, which is another important characteristic, as noted by the work [18]. To calculate the engagement metric, we pass the `hate speech-counterspeech` pair to the model, which provides a score between 0 and 1 representing the engagement in terms of upvotes/width/height. This will denote the engagement probability of that metric for the given counterspeech.

*Quality measurement metrics*: We deploy various third-party classifiers to evaluate the quality of the generated responses. To calculate the scores, we pass the generated counterspeech through the model and get the logit scores, which are passed through a softmax layer. The metrics used for evaluation are listed below.

- *Argument*: To evaluate the argument characteristic of the generated response, we use a `roberta-base-uncased` model[17] fine-tuned on the argument dataset [81]. Given this model, we pass each generated response through the classifier to predict a confidence score, which would denote the argument quality.
- *Counterargument*: In order to evaluate the counterargument characteristic of the generated response, we use a `bert-base-uncased` model trained on the counterargument dataset defined in Sect. 5.8.3. We achieve an F1-score of 0.62 on the test set of this dataset. Given this model, we pass each of the hate speech and the generated response through the classifier to predict a confidence score, which would denote the counterargument quality.
- *Counterspeech*: In order to evaluate the counterspeech quality of the generated responses, we use a `bert-base-uncased` model trained on the counterspeech dataset introduced in Sect. 5.8.3. We achieve an F1-score of 0.7 on the test set of this dataset. Given this model, we pass each generated response through the classifier to predict a confidence score, which denotes the quality of the counterspeech.
- *Toxicity*: We use the HateXplain model [20] trained on two classes—toxic and non-toxic.[18] We report the confidence between [0, 1] for the toxic class. This metric is important because a toxic counterspeech might escalate the discussion.

*Readability:* We further evaluate the readability of the counter speech generated. We use a popular readability metric Fleish Reading Ease [86] (fre). It gives a score between 0 and 100. *Type classifier*: In order to evaluate the type specific generation, we train a `bert-base-uncased` model on the type based counterspeech data points mentioned in Sect. 5.8.3 using a multi-class classification strategy. Overall, we achieve an average macro F1-score 0.80. Among the types, we achieve a macro F1-score ~0.80 for denouncing, humor, facts and affiliation. Hypocrisy is hardest to classify with an F1-score of 0.59 and questions are the easiest to classify with an F1-score of 0.97.

---

[17] https://huggingface.co/chkla/roberta-argument.
[18] https://huggingface.co/Hate-speech-CNERG/bert-base-uncased-hatexplain-rationale-two.

## 5.8.5 Results

### 5.8.5.1 Vanilla Generation
Here we discuss the evaluation results for the zero-shot evaluation of various models for the vanilla generation setting.

**Does counterspeech generation depend on model size in zero-shot setting?** We compare the small, medium (base) and large sizes of three different variations of models, i.e., DialoGPT, FlanT5 and GPT-2. We note the percentage change between the largest and smallest models. We observe that for the synthetic datasets, i.e., CONAN and CONAN-MT the change is not significant in terms of generation metrics. In terms of counterspeech quality, we see a drop of 42% for the DialoGPT model for the CONAN dataset, whereas there is a drop of 13% for GPT-2 in the case of CONAN-MT dataset. In terms of counterargument quality, we notice a drop of 6–9% across all the models except DialoGPT for CONAN dataset. Surprisingly, for CONAN-MT dataset, the toxicity increases by 44% for the GPT-2 models as we increase the model size. For the real world datasets, i.e., Gab and Reddit, there is a significant increase in toxicity (25–30%) as we increase the size of GPT-2 model. On the other hand, we find that there is an improvement in the generation quality for the DialoGPT in terms of gleu and meteor metrics. In addition, we find the readability of the counterspeech generated by DialoGPT on the Reddit dataset increases dramatically (100%) with the increase in the size of the model. The size trend for Flan-T5 is not consistent although it performs quite poorly compared to DialogGPT and GPT-2.

**Does counterspeech generation depend on model type in zero-shot setting?** We compare DialoGPT, GPT-2 and FlanT5 models since they vary in their architecture, pretraining (fine-tuning) strategies and dataset used for pre-training. In terms of synthetic datasets, FlanT5 models are better in terms of `gleu` (30%) and meteor (50%) and GPT-2 models are better in `bleurt` (8–15%). In terms of other metrics, we find that counterspeech quality is far better for GPT-2 models (200%) than other models. DialoGPT models are also better in terms of readability, but at the same time are more toxic. For the real world datasets, i.e., Gab and Reddit, DialoGPT models are better in terms of `bleurt` (7–10%), GPT-2 models are better in meteor and FlanT5 models are better in terms of `gleu` (20%). In terms of other metrics, we find that counterspeech quality and argument improve by 2–3 times for GPT-2 and FlanT5 models, but they are also higher in terms of toxicity.

### 5.8.5.2 Type Specific Generation
In this part, we evaluate the type specific generation using type prompts. We run the counterspeech classifier (described in Sect. 5.8.4) over the post generated for a particular type and measure the ratio of posts which the classifier classifies as the same type. We name this metric as `type precision`. Since we do not observe much change in the type specific generation for small, medium and large versions of the DialoGPT, FlanT5 GPT-2, we present their average performance in the Tables 5.16 and 5.17. For *affiliation*, GPT-2 and DialoGPT

**Table 5.14** Evaluation of responses generated by each model for each counterspeech generation dataset in terms of generation, engagement and quality metrics. The first column denotes which model is being used for zero-shot evaluation. DialoGPT (DGPT) and GPT-2 has s, m and l suffixes which represent 117M, 345M and 762M parameter sizes, and FlanT5 has s, b and l suffixes which represent 80M, 250M and 750M parameter sizes. For evaluating generation we measure the average gleu, meteor (met), bleurt (blrt), novelty (nov) and diversity (div). Engagement metrics consist of upvote, width, and depth. For quality, we utilise the counterspeech (cs), argument (arg), counter argument (c_arg) and toxicity (tox) scores and readability scores (fre). **Bold** denotes the best scores and higher scores denote better performance except for toxicity

| Model | gleu | met | div | nov | blrt | cs | c_arg | arg | tox (↓) | fre |
|---|---|---|---|---|---|---|---|---|---|---|
| CONAN_MT | | | | | | | | | | |
| DGPT-(s) | 0.07 | 0.08 | **0.84** | 0.84 | −1.13 | 0.15 | 0.54 | 0.26 | 0.24 | 65.21 |
| DGPT-(m) | 0.07 | 0.08 | 0.84 | 0.84 | −1.16 | 0.16 | 0.54 | 0.22 | 0.19 | **72.46** |
| DGPT-(l) | 0.07 | 0.09 | 0.83 | 0.83 | −1.14 | 0.15 | 0.59 | 0.23 | 0.28 | 66.07 |
| GPT-2 | 0.06 | 0.11 | 0.82 | 0.84 | −1.02 | 0.56 | 0.50 | 0.38 | **0.13** | 43.24 |
| GPT-2-(m) | 0.06 | 0.10 | 0.83 | **0.85** | −1.05 | 0.51 | 0.49 | 0.36 | 0.18 | 44.25 |
| GPT-2-(l) | 0.06 | 0.11 | 0.83 | 0.84 | −1.04 | 0.48 | 0.47 | 0.36 | 0.19 | 43.27 |
| flan-T5-(s) | 0.08 | 0.13 | 0.81 | 0.82 | −0.94 | 0.40 | 0.56 | 0.48 | 0.18 | 61.21 |
| flan-T5-(b) | 0.08 | 0.13 | 0.81 | 0.81 | −0.90 | 0.43 | 0.49 | 0.47 | 0.21 | 60.65 |
| flan-T5-(l) | 0.08 | 0.12 | 0.82 | 0.82 | −0.96 | 0.41 | 0.46 | 0.43 | 0.18 | 58.08 |
| ChatGPT | **0.09** | **0.17** | 0.66 | 0.80 | **−0.53** | **0.95** | **0.64** | **0.51** | 0.15 | 29.89 |
| CONAN | | | | | | | | | | |
| DGPT-(s) | 0.09 | 0.11 | **0.88** | **0.87** | −1.21 | 0.15 | 0.58 | 0.20 | 0.31 | 60.67 |
| DGPT-(m) | 0.09 | 0.11 | **0.88** | **0.87** | −1.23 | 0.09 | 0.54 | 0.15 | 0.24 | **70.95** |
| DGPT-(l) | 0.09 | 0.12 | 0.86 | 0.86 | −1.20 | 0.08 | 0.65 | 0.21 | 0.37 | 63.57 |
| GPT-2 | 0.08 | 0.15 | 0.85 | 0.86 | −1.06 | 0.48 | 0.55 | 0.37 | **0.20** | 41.09 |
| GPT-2-(m) | 0.08 | 0.15 | 0.85 | 0.86 | −1.06 | 0.34 | 0.54 | 0.38 | 0.23 | 44.65 |
| GPT-2-(l) | 0.08 | 0.15 | 0.85 | 0.86 | −1.08 | 0.43 | 0.51 | 0.35 | 0.21 | 43.73 |
| Flan-T5-(s) | 0.10 | 0.17 | 0.84 | 0.84 | −1.86 | 0.33 | 0.56 | 0.40 | 0.26 | 61.59 |
| Flan-T5-(b) | 0.10 | 0.17 | 0.84 | 0.84 | −1.84 | 0.33 | 0.52 | 0.44 | 0.25 | 53.48 |
| Flan-T5-(l) | 0.10 | 0.17 | 0.84 | 0.84 | −0.98 | 0.31 | 0.58 | 0.42 | 0.22 | 55.32 |
| ChatGPT | **0.12** | **0.23** | 0.69 | 0.81 | **−0.63** | **0.89** | **0.64** | **0.44** | 0.23 | 32.05 |
| Gab | | | | | | | | | | |
| DGPT-(s) | 0.05 | 0.07 | **0.87** | **0.86** | −1.26 | 0.06 | 0.53 | 0.06 | **0.09** | 58.18 |
| DGPT-(m) | 0.05 | 0.07 | 0.86 | **0.86** | −1.26 | 0.08 | 0.55 | 0.06 | **0.09** | 57.00 |
| DGPT-(l) | 0.05 | 0.09 | 0.85 | 0.84 | −1.28 | 0.07 | **0.56** | 0.06 | **0.09** | 59.25 |
| GPT-2 | 0.05 | 0.12 | 0.83 | 0.85 | −1.37 | 0.31 | 0.53 | 0.19 | 0.15 | 58.16 |
| GPT-2-(m) | 0.05 | 0.12 | 0.84 | 0.85 | −1.37 | 0.28 | 0.54 | 0.19 | 0.19 | 58.47 |
| GPT-2-(l) | 0.05 | 0.12 | 0.83 | 0.85 | −1.36 | 0.28 | 0.53 | 0.19 | 0.19 | 55.44 |
| FlanT5-(s) | 0.06 | 0.11 | 0.84 | 0.84 | −1.37 | 0.24 | **0.56** | 0.22 | 0.16 | 67.10 |
| FlanT5-(b) | 0.06 | 0.11 | 0.84 | 0.83 | −1.35 | 0.23 | 0.50 | 0.21 | 0.20 | **68.33** |
| FlanT5-(l) | 0.06 | 0.11 | 0.84 | 0.83 | −1.34 | 0.26 | 0.52 | 0.19 | 0.16 | 63.79 |
| ChatGPT | **0.08** | **0.17** | 0.64 | 0.80 | **−0.71** | **0.90** | 0.46 | **0.26** | 0.12 | 29.77 |

(continued)

## 5.8 LLMs for Mitigation

**Table 5.14** (continued)

| Model | gleu | met | div | nov | blrt | cs | c_arg | arg | tox (↓) | fre |
|---|---|---|---|---|---|---|---|---|---|---|
| *CONAN_MT* | | | | | | | | | | |
| *Reddit* | | | | | | | | | | |
| DGPT-(s) | 0.05 | 0.06 | **0.87** | **0.88** | −1.22 | 0.07 | 0.59 | 0.07 | **0.07** | 30.52 |
| DGPT-(m) | 0.05 | 0.07 | **0.87** | 0.87 | −1.21 | 0.08 | 0.55 | 0.06 | **0.07** | 58.24 |
| DGPT-(l) | 0.06 | 0.08 | 0.86 | 0.86 | −1.25 | 0.08 | **0.61** | 0.07 | **0.07** | 62.26 |
| GPT-2 | 0.05 | 0.12 | 0.82 | 0.86 | −1.34 | 0.36 | 0.57 | 0.21 | 0.12 | 55.06 |
| GPT-2-(m) | 0.05 | 0.12 | 0.83 | 0.86 | −1.35 | 0.35 | 0.56 | 0.22 | 0.14 | 52.91 |
| GPT-2-(l) | 0.05 | 0.12 | 0.83 | 0.86 | −1.34 | 0.35 | 0.55 | 0.21 | 0.16 | 52.88 |
| FlanT5-(s) | 0.06 | 0.12 | 0.83 | 0.84 | −1.35 | 0.31 | 0.57 | 0.26 | 0.12 | **73.82** |
| FlanT5-(b) | 0.06 | 0.11 | 0.84 | 0.84 | −1.34 | 0.29 | 0.51 | 0.22 | 0.16 | 70.51 |
| FlanT5-(l) | 0.06 | 0.11 | 0.84 | 0.84 | −1.32 | 0.34 | 0.53 | 0.20 | 0.11 | 70.99 |
| ChatGPT | **0.08** | **0.17** | 0.67 | 0.81 | **−0.77** | **0.85** | 0.50 | **0.26** | 0.13 | 29.12 |

**Table 5.15** Counterspeech generation by different models in a zero-shot setting for vanilla generation. The posts are taken from one of the available datasets. For all models except ChatGPT, we use the `l` version of the models

| Model | Hate speech | Counterspeech |
|---|---|---|
| FlanT5 | Muslims have no place in civilised Europe. They should go home | We shall not discriminate on the grounds of race, religion, colour, culture or national origin. Our society defends and protects its own dignity |
| DGPT | We are being invaded by Muslims. They are taking over our countries | lol s I hope we all get this reference. The irony is lost at the end of that sentence isn't it... |
| ChatGPT | If you are a Jew you are for sure involved in 9/11 planning. This is all about Race and 'Enlightenment'. You are all guilty until proven innocent | It is unfair and untrue to suggest that all Jews were involved in the planning of 9/11... |

perform better with cluster-centered prompts across all the datasets, while FlanT5 performs better with frequency-based prompts. Cluster centered prompts improve the baseline type precision by 0.41 and 0.32 units for GPT-2 and DialoGPT respectively and frequency-based prompts improve the FlanT5 by 0.16 units. For *denouncing*, GPT-2, DialoGPT and FlanT5 perform better with manual prompts. The type precision of FlanT5 increases by 0.16 units while the other models increase by 0.36 units. *Facts* type[19] are one of the categories where the GPT-2 and FlanT5 performs quite better than DialoGPT type precision. Overall, the baseline FlanT5 model is also good for generating fact type statements for CONAN and CONAN-MT datasets, with scoring ~0.7 in type precision. Here, also manual prompts are

---

[19] We don't claim that these models are generating factual statements, but fact-styled statements.

**Table 5.16** Evaluation of responses generated by each model and prompt strategy. The first column denotes which model is being used for zero-shot evaluation. DialoGPT, GPT-2 and FlanT5 are averaged across three parameter sizes. The second column denotes the prompt strategy out of manual_prompt (manual), frequency-based (freq), cluster-centred (cluster) being used, where baseline (base) represents no prompt strategy. The next six columns represent the type-precision for each model + `type_prompt`. aff: affiliation, den: denouncing, fac: facts, hum: humour, hyp: hypocrisy, qu: question. **Bold** denotes the best scores and higher scores denote better performance

| Model | prompt | aff | den | fac | hum | hyp | qu |
|---|---|---|---|---|---|---|---|
| *CONAN_MT* | | | | | | | |
| GPT-2 | base | 0.04 | 0.04 | 0.60 | 0.03 | 0.29 | 0.00 |
|  | manual | 0.06 | **0.40** | **0.78** | **0.20** | 0.21 | 0.01 |
|  | freq | 0.06 | 0.07 | 0.77 | 0.10 | 0.21 | **0.05** |
|  | cluster | **0.46** | 0.34 | 0.72 | 0.07 | **0.43** | 0.03 |
| DGPT | base | 0.04 | 0.10 | 0.29 | 0.19 | 0.39 | 0.00 |
|  | manual | 0.10 | **0.45** | **0.46** | **0.46** | **0.51** | 0.01 |
|  | freq | 0.10 | 0.15 | 0.44 | 0.34 | 0.50 | **0.08** |
|  | cluster | **0.38** | 0.41 | 0.42 | 0.29 | 0.49 | 0.03 |
| FlanT5 | base | 0.04 | 0.06 | 0.73 | 0.03 | 0.14 | 0.00 |
|  | manual | 0.15 | **0.23** | **0.81** | **0.15** | 0.10 | 0.00 |
|  | freq | **0.26** | 0.06 | 0.80 | 0.07 | 0.21 | 0.00 |
|  | cluster | 0.20 | 0.18 | 0.74 | 0.10 | **0.23** | 0.00 |
| ChatGPT | base | 0.02 | 0.22 | 0.75 | 0.00 | 0.01 | 0.00 |
|  | manual | 0.03 | 0.40 | 0.81 | 0.00 | 0.02 | 0.00 |
|  | freq | **0.51** | 0.31 | **0.94** | 0.00 | 0.06 | 0.01 |
|  | cluster | 0.27 | **0.46** | 0.77 | 0.00 | **0.09** | **0.01** |
| *CONAN* | | | | | | | |
| GPT-2 | base | 0.02 | 0.04 | 0.65 | 0.01 | 0.27 | 0.00 |
|  | manual | 0.05 | **0.37** | **0.81** | **0.13** | 0.19 | 0.01 |
|  | freq | 0.04 | 0.07 | 0.79 | 0.05 | 0.18 | **0.04** |
|  | cluster | **0.44** | 0.33 | 0.76 | 0.04 | **0.39** | 0.02 |
| DGPT | base | 0.02 | 0.09 | 0.28 | 0.20 | 0.41 | 0.00 |
|  | manual | 0.07 | **0.44** | **0.42** | **0.46** | **0.52** | 0.01 |
|  | freq | 0.07 | 0.16 | 0.42 | 0.37 | 0.50 | **0.07** |
|  | cluster | **0.36** | 0.41 | 0.35 | 0.32 | 0.50 | 0.03 |
| FlanT5 | base | 0.02 | 0.08 | 0.75 | 0.02 | 0.12 | 0.00 |
|  | manual | 0.10 | **0.23** | 0.79 | **0.14** | 0.12 | 0.00 |
|  | freq | **0.18** | 0.07 | **0.80** | 0.05 | 0.21 | 0.00 |
|  | cluster | 0.16 | 0.17 | 0.73 | 0.08 | **0.23** | 0.00 |
| ChatGPT | base | 0.04 | **0.64** | 0.28 | 0.00 | **0.10** | 0.00 |
|  | manual | 0.02 | 0.39 | 0.93 | 0.00 | 0.02 | 0.00 |
|  | freq | **0.52** | 0.30 | **0.96** | 0.00 | 0.06 | 0.00 |
|  | cluster | 0.26 | 0.43 | 0.87 | **0.01** | 0.10 | **0.01** |

the best, and they improve the baseline type precision by 0.20, 0.15 and 0.13 units for GPT-2, DialoGPT and FlanT5 respectively. For *humour*, we find again that the manual prompts are

## 5.8 LLMs for Mitigation

**Table 5.17** Evaluation of responses generated by each model and prompt strategy. The first column denotes which model is being used for zero-shot evaluation. DialoGPT, GPT-2 and FlanT5 are averaged across three parameter sizes. The second column denotes the prompt strategy out of manual_prompt (manual), frequency based (freq), cluster centered (cluster) being used, where baseline (base) represents no prompt strategy. The next six columns represent the type-precision for each model + `type_prompt`. aff: affiliation, den: denouncing, fac: facts, hum: humour, hyp: hypocrisy, qu: question. **Bold** denotes the best scores and higher scores denote better performance

| Model | prompt | aff | den | fac | hum | hyp | qu |
|---|---|---|---|---|---|---|---|
| *Reddit* | | | | | | | |
| GPT-2 | base | 0.08 | 0.07 | 0.27 | 0.17 | 0.41 | 0.00 |
| | manual | 0.20 | **0.48** | **0.51** | **0.40** | 0.48 | 0.01 |
| | freq | 0.20 | 0.16 | 0.50 | 0.26 | 0.50 | **0.08** |
| | cluster | **0.47** | 0.41 | 0.47 | 0.21 | **0.50** | 0.03 |
| DGPT | base | 0.03 | 0.07 | 0.14 | 0.47 | 0.30 | 0.00 |
| | manual | 0.08 | **0.43** | **0.30** | **0.75** | **0.52** | 0.01 |
| | freq | 0.08 | 0.14 | 0.29 | 0.64 | 0.52 | **0.07** |
| | cluster | **0.33** | 0.38 | 0.29 | 0.52 | 0.41 | 0.02 |
| FlanT5 | base | 0.09 | 0.12 | 0.26 | 0.23 | 0.30 | 0.00 |
| | manual | 0.17 | **0.27** | **0.44** | **0.34** | 0.30 | 0.00 |
| | freq | **0.23** | 0.12 | 0.41 | 0.25 | 0.30 | **0.01** |
| | cluster | 0.19 | 0.21 | 0.43 | 0.24 | **0.31** | **0.01** |
| ChatGPT | base | 0.07 | 0.39 | 0.44 | 0.00 | 0.10 | 0.00 |
| | manual | 0.07 | **0.60** | **0.75** | 0.02 | 0.10 | 0.00 |
| | freq | **0.54** | 0.48 | 0.75 | 0.00 | **0.13** | 0.01 |
| | cluster | 0.35 | 0.57 | 0.57 | 0.01 | 0.12 | **0.01** |
| *Gab* | | | | | | | |
| GPT-2 | base | 0.08 | 0.09 | 0.26 | 0.18 | 0.40 | 0.00 |
| | manual | 0.23 | **0.49** | **0.51** | **0.40** | 0.47 | 0.01 |
| | freq | 0.23 | 0.19 | 0.48 | 0.28 | 0.46 | **0.07** |
| | cluster | **0.50** | 0.44 | 0.45 | 0.20 | **0.49** | 0.03 |
| DGPT | base | 0.03 | 0.08 | 0.12 | 0.47 | 0.29 | 0.00 |
| | manual | 0.09 | **0.46** | **0.27** | **0.73** | 0.52 | 0.01 |
| | freq | 0.09 | 0.16 | 0.27 | 0.64 | **0.53** | **0.07** |
| | cluster | **0.33** | 0.41 | 0.26 | 0.52 | 0.39 | 0.02 |
| FlanT5 | base | 0.10 | 0.12 | 0.24 | 0.25 | 0.29 | 0.00 |
| | manual | 0.17 | **0.28** | **0.43** | **0.37** | 0.29 | 0.00 |
| | freq | **0.23** | 0.11 | 0.41 | 0.28 | 0.29 | 0.00 |
| | cluster | 0.19 | 0.23 | **0.43** | 0.26 | **0.33** | 0.00 |
| ChatGPT | base | 0.02 | 0.09 | **0.88** | 0.00 | 0.01 | 0.00 |
| | manual | 0.07 | **0.71** | 0.62 | **0.01** | 0.07 | 0.00 |
| | freq | **0.52** | 0.66 | 0.65 | 0.00 | **0.09** | 0.00 |
| | cluster | 0.30 | 0.69 | 0.52 | 0.00 | 0.07 | 0.00 |

the best. They improve the baseline type precision by 0.185, 0.27 and 0.12 units for GPT-2, DialoGPT and FlanT5 respectively. For *hypocrisy* and *questions*, there is no one prompting

**Table 5.18** Counterspeech generated by different models in a zero-shot setting for type-specific generation where the type is denoted in the brackets. For each type, we select the prompt strategy based on the performance of that model on that type. aff: affiliation, den: denouncing, fac: facts, hum: humour, hyp: hypocrisy, qu: question. For all models except ChatGPT, we use their larger versions

| Model (Type) | Hate speech | Counterspeech |
|---|---|---|
| GPT-2 (aff) | Blacks are just an inferior bloodline to our own. We should remove them | I am muslim but my black brothers don't respect us as true Muslims, and there is another cause of that.. so how do I make their hearts be happy then they can accept me? the one thing we have in common with other races....is faith—you know all over this world...every people has a belief" |
| GPT-2 (den) | No, only the Severely Retarded get the job | I think maybe prejudice against people with that disease does drive a lot of discrimination in their lives—they've got some serious issues and it's very hard to help them |
| DialoGPT (hum) | Never said it was, faggot | This is funny. uhh you know who loves the attention and self hate speech as well? Donald Trump, |
| DialoGPT (hyp) | People of a different race, including transgenders, are very few. With some effort we can contain them. | In contradiction to the people living in fear and ignorance. gender neutrality exists even though those who want it aren't happy about their pronouns? |
| FlanT5 (aff) | We should expel Muslims, everyone wants them out of the country. | I have no issue with Muslims. I just disagree that they are the bad guys and there are lots of other countries like my country in which they won't go, so we shouldn't do anything about them. |
| ChatGPT (facts) | Women should not be allowed to vote. They are subhuman. | I am a Muslim and in our faith, both men and women are valued equally. Discriminating against one gender is unjustified, as it goes against the principles of justice that Islam teaches us to follow.. |

strategy that works for both the models and all the datasets. Cluster centered prompts are better for GPT-2 (0.10 units) and manual prompts are better for DialoGPT (0.15 units). Surprisingly, none of the models reach 0.1 type precision for the *question* type. We have added some examples of type specific generation in the Table 5.18.

### 5.8.5.3 Generation Using ChatGPT

In this part, we look at the counterspeech generated by ChatGPT. In terms of vanilla generation, as noted in Table 5.14, we notice that ChatGPT performs better than other models in terms of generation quality. It improves the gleu (12%), meteor (32%) and bleurt (42.25%). Among other metrics, ChatGPT improves the counterspeech quality by 120% and improves the argument quality by 27%. The toxicity scores are comparable, although they are slightly higher than the best models. One interesting point is that readability of the ChatGPT texts reduces by ~35%. We also note some counterspeech generated by ChatGPT and compare them with the other models in Table 5.15. In terms of type specific generation as noted in Tables 5.16 and 5.17, we find that for *affiliation*, frequency based prompts improves the baseline type precision by 0.49 units across all datasets. Other than that, other types do not have a consistent best prompt strategy that works across all the categories. Interestingly, for some cases like denouncing type for CONAN, the model performance worsens if we introduce any prompt. Further ChatGPT model performs best in fact-type counterspeech scoring close to 0.9 in three out of four datasets. ChatGPT struggles for the types—humour, hypocrisy, and questions even in presence of type prompts and their type precisions rarely reach above 0.1 across datasets (Table 5.17).

# References

1. Wikipedia contributors. *Hate Speech Laws in India — Wikipedia, The Free Encyclopedia*, 2024. [Online; accessed 22-September-2024].
2. Wikipedia contributors. *Hate Speech in the United States — Wikipedia, The Free Encyclopedia*, 2024. [Online; accessed 22-September-2024].
3. Paula Fortuna and Sérgio Nunes. A survey on automatic detection of hate speech in text. *ACM Comput. Surv.*, 51(4), July 2018a. ISSN 0360-0300.
4. Sarah T Roberts. *Behind the screen: The hidden digital labor of commercial content moderation*. University of Illinois at Urbana-Champaign, 2014.
5. Eshwar Chandrasekharan, Umashanthi Pavalanathan, Anirudh Srinivasan, Adam Glynn, Jacob Eisenstein, and Eric Gilbert. You can't stay here: The efficacy of reddit's 2015 ban examined through hate speech. *Proc. ACM Hum.-Comput. Interact.*, 1(CSCW), December 2017.
6. Shiza Ali, Mohammad Hammas Saeed, Esraa Aldreabi, Jeremy Blackburn, Emiliano De Cristofaro, Savvas Zannettou, and Gianluca Stringhini. Understanding the effect of deplatforming on social networks. 2021.
7. Raj Ratn Pranesh, Ambesh Shekhar, and Anish Kumar. Towards automatic online hate speech intervention generation using pretrained language model. 2021.
8. Ziwei Ji, Nayeon Lee, Rita Frieske, Tiezheng Yu, Dan Su, Yan Xu, Etsuko Ishii, Ye Jin Bang, Andrea Madotto, and Pascale Fung. Survey of hallucination in natural language generation. *ACM Computing Surveys*, 55(12):1–38, 2023.
9. Jing Qian, Anna Bethke, Yinyin Liu, Elizabeth Belding, and William Yang Wang. A benchmark dataset for learning to intervene in online hate speech. In *Proceedings of (EMNLP-IJCNLP)*, pages 4755–4764, November 2019a.
10. Binny Mathew, Punyajoy Saha, Hardik Tharad, Subham Rajgaria, Prajwal Singhania, Suman Kalyan Maity, Pawan Goyal, and Animesh Mukherjee. Thou shalt not hate: Countering online hate speech. In *Proceedings of the international AAAI conference on web and social media*, volume 13, pages 369–380, 2019b.
11. Stefanie Ullmann and Marcus Tomalin. Quarantining online hate speech: technical and ethical perspectives. *Ethics and Information Technology*, 22:69–80, 2020.
12. Corrado Fumagalli. Counterspeech and ordinary citizens: how? when? *Political Theory*, 49(6):1021–1047, 2021.

13. Bertie Vidgen, Helen Margetts, and Alex Harris. How much online abuse is there. *Alan Turing Institute*, 2019.
14. Susan Benesch, Derek Ruths, Kelly P Dillon, Haji Mohammad Saleem, and Lucas Wright. Counterspeech on twitter: A field study. *Dangerous Speech Project. Available at:* https://dangerousspeech.org/counterspeech-on-twitter-a-field-study/, 2016b.
15. Jean-Leon Beauvois, Robert-Vincent Joule, and Fabien Brunetti. Cognitive rationalization and act rationalization in an escalation of commitment. *Basic and applied social psychology*, 14(1):1–17, 1993.
16. Susan Benesch. Countering dangerous speech: New ideas for genocide prevention. *Washington, DC: United States Holocaust Memorial Museum*, 2014.
17. Ke Wang and Xiaojun Wan. Sentigan: Generating sentimental texts via mixture adversarial networks. In *IJCAI*, pages 4446–4452, 2018.
18. Susan Benesch, Derek Ruths, Kelly P Dillon, Haji Mohammad Saleem, and Lucas Wright. Considerations for successful counterspeech. *A report for Public Safety Canada under the Kanishka Project. Accessed November*, 25:2020, 2016a.
19. Xiang Gao, Yizhe Zhang, Michel Galley, Chris Brockett, and Bill Dolan. Dialogue response ranking training with large-scale human feedback data. In Bonnie Webber, Trevor Cohn, Yulan He, and Yang Liu, editors, *Proceedings of the 2020 Conference on Empirical Methods in Natural Language Processing (EMNLP)*, pages 386–395, Online, November 2020. Association for Computational Linguistics. URL https://aclanthology.org/2020.emnlp-main.28.
20. Binny Mathew, Punyajoy Saha, Seid Muhie Yimam, Chris Biemann, Pawan Goyal, and Animesh Mukherjee. Hatexplain: A benchmark dataset for explainable hate speech detection. *arXiv preprint* arXiv:2012.10289, 2020b.
21. Serra Sinem Tekiroglu, Helena Bonaldi, Margherita Fanton, and Marco Guerini. Using pretrained language models for producing counter narratives against hate speech: a comparative study. *arXiv preprint* arXiv:2204.01440, 2022.
22. Margherita Fanton, Helena Bonaldi, Serra Sinem Tekiroğlu, and Marco Guerini. Human-in-the-loop for data collection: a multi-target counter narrative dataset to fight online hate speech. In *Proceedings of the 59th Annual Meeting of the Association for Computational Linguistics and the 11th International Joint Conference on Natural Language Processing (Volume 1: Long Papers)*, pages 3226–3240, Online, August 2021. Association for Computational Linguistics. URL https://aclanthology.org/2021.acl-long.250.
23. Jacob Devlin, Ming-Wei Chang, Kenton Lee, and Kristina Toutanova. BERT: Pre-training of deep bidirectional transformers for language understanding. In Jill Burstein, Christy Doran, and Thamar Solorio, editors, *Proceedings of the 2019 Conference of the North American Chapter of the Association for Computational Linguistics: Human Language Technologies, Volume 1 (Long and Short Papers)*, pages 4171–4186, Minneapolis, Minnesota, June 2019. Association for Computational Linguistics. URL https://aclanthology.org/N19-1423.
24. Alex Wang and Kyunghyun Cho. Bert has a mouth, and it must speak: Bert as a markov random field language model. In *Proceedings of the Workshop on Methods for Optimizing and Evaluating Neural Language Generation*, pages 30–36, 2019.
25. Sascha Rothe, Shashi Narayan, and Aliaksei Severyn. Leveraging pre-trained checkpoints for sequence generation tasks. *Transactions of the Association for Computational Linguistics*, 8:264–280, 2020. URL https://aclanthology.org/2020.tacl-1.18.
26. Alec Radford, Jeff Wu, Rewon Child, David Luan, Dario Amodei, and Ilya Sutskever. Language models are unsupervised multitask learners. 2019.
27. Yizhe Zhang, Siqi Sun, Michel Galley, Yen-Chun Chen, Chris Brockett, Xiang Gao, Jianfeng Gao, Jingjing Liu, and Bill Dolan. DIALOGPT : Large-scale generative pre-training for conversational response generation. In Asli Celikyilmaz and Tsung-Hsien Wen, editors, *Proceedings of the*

*58th Annual Meeting of the Association for Computational Linguistics: System Demonstrations*, pages 270–278, Online, July 2020b. Association for Computational Linguistics. URL https://aclanthology.org/2020.acl-demos.30.

28. Mike Lewis, Yinhan Liu, Naman Goyal, Marjan Ghazvininejad, Abdelrahman Mohamed, Omer Levy, Veselin Stoyanov, and Luke Zettlemoyer. BART: Denoising sequence-to-sequence pre-training for natural language generation, translation, and comprehension. In Dan Jurafsky, Joyce Chai, Natalie Schluter, and Joel Tetreault, editors, *Proceedings of the 58th Annual Meeting of the Association for Computational Linguistics*, pages 7871–7880, Online, July 2020. Association for Computational Linguistics. URL https://aclanthology.org/2020.acl-main.703.

29. Colin Raffel, Noam Shazeer, Adam Roberts, Katherine Lee, Sharan Narang, Michael Matena, Yanqi Zhou, Wei Li, and Peter J Liu. Exploring the limits of transfer learning with a unified text-to-text transformer. *The Journal of Machine Learning Research*, 21(1):5485–5551, 2020b.

30. Jiwei Li, Will Monroe, Alan Ritter, Dan Jurafsky, Michel Galley, and Jianfeng Gao. Deep reinforcement learning for dialogue generation. In Jian Su, Kevin Duh, and Xavier Carreras, editors, *Proceedings of the 2016 Conference on Empirical Methods in Natural Language Processing*, pages 1192–1202, Austin, Texas, November 2016b. Association for Computational Linguistics. URL https://aclanthology.org/D16-1127.

31. Sam Wiseman, Stuart Shieber, and Alexander Rush. Challenges in data-to-document generation. In Martha Palmer, Rebecca Hwa, and Sebastian Riedel, editors, *Proceedings of the 2017 Conference on Empirical Methods in Natural Language Processing*, pages 2253–2263, Copenhagen, Denmark, September 2017. Association for Computational Linguistics. URL https://aclanthology.org/D17-1239.

32. Angela Fan, Mike Lewis, and Yann Dauphin. Hierarchical neural story generation. In Iryna Gurevych and Yusuke Miyao, editors, *Proceedings of the 56th Annual Meeting of the Association for Computational Linguistics (Volume 1: Long Papers)*, pages 889–898, Melbourne, Australia, July 2018. Association for Computational Linguistics. URL https://aclanthology.org/P18-1082.

33. Ari Holtzman, Jan Buys, Li Du, Maxwell Forbes, and Yejin Choi. The curious case of neural text degeneration. In *International Conference on Learning Representations*, 2020. URL https://openreview.net/forum?id=rygGQyrFvH.

34. Wanzheng Zhu and Suma Bhat. Generate, prune, select: A pipeline for counterspeech generation against online hate speech. In *Findings of the Association for Computational Linguistics: ACL-IJCNLP 2021*, pages 134–149, Online, August 2021. Association for Computational Linguistics. URL https://aclanthology.org/2021.findings-acl.12.

35. Samuel R. Bowman, Luke Vilnis, Oriol Vinyals, Andrew Dai, Rafal Jozefowicz, and Samy Bengio. Generating sentences from a continuous space. In Stefan Riezler and Yoav Goldberg, editors, *Proceedings of the 20th SIGNLL Conference on Computational Natural Language Learning*, pages 10–21, Berlin, Germany, August 2016. Association for Computational Linguistics. URL https://aclanthology.org/K16-1002.

36. Matthew Henderson, Ivan Vulić, Daniela Gerz, Iñigo Casanueva, Paweł Budzianowski, Sam Coope, Georgios Spithourakis, Tsung-Hsien Wen, Nikola Mrkšić, and Pei-Hao Su. Training neural response selection for task-oriented dialogue systems. *arXiv preprint* arXiv:1906.01543, 2019.

37. Xiang Gao, Sungjin Lee, Yizhe Zhang, Chris Brockett, Michel Galley, Jianfeng Gao, and Bill Dolan. Jointly optimizing diversity and relevance in neural response generation. In Jill Burstein, Christy Doran, and Thamar Solorio, editors, *Proceedings of the 2019 Conference of the North American Chapter of the Association for Computational Linguistics: Human Language Technologies, Volume 1 (Long and Short Papers)*, pages 1229–1238, Minneapolis, Minnesota, June 2019. Association for Computational Linguistics. URL https://aclanthology.org/N19-1125.

38. Yi-Ling Chung, Elizaveta Kuzmenko, Serra Sinem Tekiroglu, and Marco Guerini. CONAN - COunter NArratives through nichesourcing: a multilingual dataset of responses to fight online hate speech. In *Proceedings of the 57th Annual Meeting of the Association for Computational Linguistics*, pages 2819–2829, Florence, Italy, July 2019. Association for Computational Linguistics. URL https://www.aclweb.org/anthology/P19-1271.
39. Ilya Sutskever, Oriol Vinyals, and Quoc V Le. Sequence to sequence learning with neural networks. *Advances in neural information processing systems*, 27, 2014.
40. Jiwei Li, Michel Galley, Chris Brockett, Jianfeng Gao, and Bill Dolan. A diversity-promoting objective function for neural conversation models. In Kevin Knight, Ani Nenkova, and Owen Rambow, editors, *Proceedings of the 2016 Conference of the North American Chapter of the Association for Computational Linguistics: Human Language Technologies*, pages 110–119, San Diego, California, June 2016a. Association for Computational Linguistics. URL https://aclanthology.org/N16-1014.
41. Marc'Aurelio Ranzato, Sumit Chopra, Michael Auli, and Wojciech Zaremba. Sequence level training with recurrent neural networks. *arXiv preprint* arXiv:1511.06732, 2015.
42. Yizhe Zhang, Michel Galley, Jianfeng Gao, Zhe Gan, Xiujun Li, Chris Brockett, and Bill Dolan. Generating informative and diverse conversational responses via adversarial information maximization. In S. Bengio, H. Wallach, H. Larochelle, K. Grauman, N. Cesa-Bianchi, and R. Garnett, editors, *Advances in Neural Information Processing Systems*, volume 31. Curran Associates, Inc., 2018.
43. Yaoming Zhu, Sidi Lu, Lei Zheng, Jiaxian Guo, Weinan Zhang, Jun Wang, and Yong Yu. Texygen: A benchmarking platform for text generation models. In *The 41st International ACM SIGIR Conference on Research & Development in Information Retrieval*, pages 1097–1100, 2018.
44. Kishore Papineni, Salim Roukos, Todd Ward, and Wei-Jing Zhu. Bleu: a method for automatic evaluation of machine translation. In Pierre Isabelle, Eugene Charniak, and Dekang Lin, editors, *Proceedings of the 40th Annual Meeting of the Association for Computational Linguistics*, pages 311–318, Philadelphia, Pennsylvania, USA, July 2002a. Association for Computational Linguistics. URL https://aclanthology.org/P02-1040.
45. Chin-Yew Lin and Eduard Hovy. Automatic evaluation of summaries using n-gram co-occurrence statistics. In *Proceedings of the 2003 Human Language Technology Conference of the North American Chapter of the Association for Computational Linguistics*, pages 150–157, 2003. URL https://aclanthology.org/N03-1020.
46. Wei Zhao, Maxime Peyrard, Fei Liu, Yang Gao, Christian M. Meyer, and Steffen Eger. MoverScore: Text generation evaluating with contextualized embeddings and earth mover distance. In Kentaro Inui, Jing Jiang, Vincent Ng, and Xiaojun Wan, editors, *Proceedings of the 2019 Conference on Empirical Methods in Natural Language Processing and the 9th International Joint Conference on Natural Language Processing (EMNLP-IJCNLP)*, pages 563–578, Hong Kong, China, November 2019. Association for Computational Linguistics. URL https://aclanthology.org/D19-1053.
47. Tianyi Zhang, Varsha Kishore, Felix Wu, Kilian Q Weinberger, and Yoav Artzi. Bertscore: Evaluating text generation with bert. *arXiv preprint* arXiv:1904.09675, 2019b.
48. Wanzheng Zhu and Suma Bhat. Gruen for evaluating linguistic quality of generated text. *arXiv preprint* arXiv:2010.02498, 2020.
49. Jamie Bartlett and Alex Krasodomski-Jones. Counter-speech examining content that challenges extremism online. *DEMOS, October*, 2015.
50. Ben Krause, Akhilesh Deepak Gotmare, Bryan McCann, Nitish Shirish Keskar, Shafiq Joty, Richard Socher, and Nazneen Fatema Rajani. Gedi: Generative discriminator guided sequence generation. *arXiv preprint* arXiv:2009.06367, 2020.

51. Punyajoy Saha, Kanishk Singh, Adarsh Kumar, Binny Mathew, and Animesh Mukherjee. Countergedi: A controllable approach to generate polite, detoxified and emotional counterspeech. In Lud De Raedt, editor, *Proceedings of the Thirty-First International Joint Conference on Artificial Intelligence, IJCAI-22*, pages 5157–5163. International Joint Conferences on Artificial Intelligence Organization, 7 2022. AI for Good.
52. Leigh Clark, Nadia Pantidi, Orla Cooney, Philip Doyle, Diego Garaialde, Justin Edwards, Brendan Spillane, Emer Gilmartin, Christine Murad, Cosmin Munteanu, Vincent Wade, and Benjamin R. Cowan. *What Makes a Good Conversation? Challenges in Designing Truly Conversational Agents*, page 1-12. Association for Computing Machinery, New York, NY, USA, 2019. ISBN 9781450359702.
53. Aman Madaan, Amrith Setlur, Tanmay Parekh, Barnabas Poczos, Graham Neubig, Yiming Yang, Ruslan Salakhutdinov, Alan W Black, and Shrimai Prabhumoye. Politeness transfer: A tag and generate approach. In Dan Jurafsky, Joyce Chai, Natalie Schluter, and Joel Tetreault, editors, *Proceedings of the 58th Annual Meeting of the Association for Computational Linguistics*, pages 1869–1881, Online, July 2020. Association for Computational Linguistics. URL https://aclanthology.org/2020.acl-main.169.
54. Helmut Prendinger and Mitsuru Ishizuka. The empathic companion: A character-based interface that addresses users' affective states. *App. AI*, 2005.
55. Elvis Saravia, Hsien-Chi Toby Liu, Yen-Hao Huang, Junlin Wu, and Yi-Shin Chen. Carer: Contextualized affect representations for emotion recognition. In *EMNLP*, 2018.
56. Ananya B Sai, Akash Kumar Mohankumar, and Mitesh M Khapra. A survey of evaluation metrics used for nlg systems. *arXiv preprint* arXiv:2008.12009, 2020.
57. Alex Warstadt and Samuel R Bowman. Linguistic analysis of pretrained sentence encoders with acceptability judgments. *arXiv preprint* arXiv:1901.03438, 2019.
58. Serra Sinem Tekiroğlu, Yi-Ling Chung, and Marco Guerini. Generating counter narratives against online hate speech: Data and strategies. In *Proceedings of the 58th Annual Meeting of the Association for Computational Linguistics*, pages 1177–1190, 2020.
59. Yi-Ling Chung, Serra Sinem Tekiroğlu, and Marco Guerini. Towards knowledge-grounded counter narrative generation for hate speech. In Chengqing Zong, Fei Xia, Wenjie Li, and Roberto Navigli, editors, *Findings of the Association for Computational Linguistics: ACL-IJCNLP 2021*, pages 899–914, Online, August 2021b. Association for Computational Linguistics. URL https://aclanthology.org/2021.findings-acl.79.
60. Max Grusky, Mor Naaman, and Yoav Artzi. Newsroom: A dataset of 1.3 million summaries with diverse extractive strategies. In Marilyn Walker, Heng Ji, and Amanda Stent, editors, *Proceedings of the 2018 Conference of the North American Chapter of the Association for Computational Linguistics: Human Language Technologies, Volume 1 (Long Papers)*, pages 708–719, New Orleans, Louisiana, June 2018. Association for Computational Linguistics. URL https://aclanthology.org/N18-1065.
61. Stephen Merity, Caiming Xiong, James Bradbury, and Richard Socher. Pointer sentinel mixture models. *arXiv preprint* arXiv:1609.07843, 2016.
62. Peter D Turney. Learning algorithms for keyphrase extraction. *Information retrieval*, 2:303–336, 2000.
63. Giovanni Moretti, Rachele Sprugnoli, Sara Tonelli, et al. Digging in the dirt: Extracting keyphrases from texts with kd. *CLiC it*, 198, 2015.
64. Xinyu Hua, Zhe Hu, and Lu Wang. Argument generation with retrieval, planning, and realization. *arXiv preprint* arXiv:1906.03717, 2019.
65. Ashish Vaswani, Noam Shazeer, Niki Parmar, Jakob Uszkoreit, Llion Jones, Aidan N Gomez, Łukasz Kaiser, and Illia Polosukhin. Attention is all you need. *Advances in neural information processing systems*, 30, 2017.

66. Stephen E Robertson, Steve Walker, Susan Jones, Micheline M Hancock-Beaulieu, Mike Gatford, et al. Okapi at trec-3. *Nist Special Publication Sp*, 109:109, 1995.
67. Jingqing Zhang, Yao Zhao, Mohammad Saleh, and Peter Liu. Pegasus: Pre-training with extracted gap-sentences for abstractive summarization. In *International Conference on Machine Learning*, pages 11328–11339. PMLR, 2020a.
68. Fabio Petroni, Tim Rocktäschel, Patrick Lewis, Anton Bakhtin, Yuxiang Wu, Alexander H Miller, and Sebastian Riedel. Language models as knowledge bases? *arXiv preprint* arXiv:1909.01066, 2019.
69. Zewen Chi, Li Dong, Furu Wei, Wenhui Wang, Xian-Ling Mao, and Heyan Huang. Cross-lingual natural language generation via pre-training. In *Proceedings of the AAAI conference on artificial intelligence*, volume 34, pages 7570–7577, 2020.
70. Mauro Cettolo, Nicola Bertoldi, and Marcello Federico. The repetition rate of text as a predictor of the effectiveness of machine translation adaptation. In *Proceedings of the 11th Biennial Conference of the Association for Machine Translation in the Americas (AMTA 2014)*, pages 166–179, 2014. URL http://www.mt-archive.info/10/AMTA-2014-Cettolo.pdf.
71. Jing Qian, Anna Bethke, Yinyin Liu, Elizabeth Belding, and William Yang Wang. A benchmark dataset for learning to intervene in online hate speech. In *Proceedings of the 2019 Conference on Empirical Methods in Natural Language Processing and the 9th International Joint Conference on Natural Language Processing (EMNLP-IJCNLP)*, pages 4757–4766, 2019b.
72. Hyung Won Chung, Le Hou, Shayne Longpre, Barret Zoph, Yi Tay, William Fedus, Yunxuan Li, Xuezhi Wang, Mostafa Dehghani, Siddhartha Brahma, et al. Scaling instruction-finetuned language models. *arXiv preprint* arXiv:2210.11416, 2022.
73. Colin Raffel, Noam Shazeer, Adam Roberts, Katherine Lee, Sharan Narang, Michael Matena, Yanqi Zhou, Wei Li, and Peter J. Liu. Exploring the limits of transfer learning with a unified text-to-text transformer. *Journal of Machine Learning Research*, 21(140):1–67, 2020a. URL http://jmlr.org/papers/v21/20-074.html.
74. Long Ouyang, Jeffrey Wu, Xu Jiang, Diogo Almeida, Carroll Wainwright, Pamela Mishkin, Chong Zhang, Sandhini Agarwal, Katarina Slama, Alex Ray, et al. Training language models to follow instructions with human feedback. *Advances in Neural Information Processing Systems*, 35:27730–27744, 2022.
75. Jason Wei, Xuezhi Wang, Dale Schuurmans, Maarten Bosma, Fei Xia, Ed Chi, Quoc V Le, Denny Zhou, et al. Chain-of-thought prompting elicits reasoning in large language models. *Advances in Neural Information Processing Systems*, 35:24824–24837, 2022.
76. OpenAI. Introducing chatgpt. https://openai.com/blog/chatgpt, November 2022. (Accessed on 06/11/2023).
77. Jingfeng Yang, Hongye Jin, Ruixiang Tang, Xiaotian Han, Qizhang Feng, Haoming Jiang, Shaochen Zhong, Bing Yin, and Xia Hu. Harnessing the power of llms in practice: A survey on chatgpt and beyond. *ACM Trans. Knowl. Discov. Data*, 18(6), April 2024. ISSN 1556-4681. URL https://doi.org/10.1145/3649506.
78. Dhendra Marutho, Sunarna Hendra Handaka, Ekaprana Wijaya, and Muljono. The determination of cluster number at k-mean using elbow method and purity evaluation on headline news. In *2018 International Seminar on Application for Technology of Information and Communication*, pages 533–538, 2018.
79. Yi-Ling Chung, Marco Guerini, and Rodrigo Agerri. Multilingual counter narrative type classification. *arXiv preprint* arXiv:2109.13664, 2021a.
80. Yi-Ling Chung, Serra Sinem Tekiroğlu, and Marco Guerini. Towards knowledge-grounded counter narrative generation for hate speech. In *Findings of the Association for Computational Linguistics: ACL-IJCNLP 2021*, pages 899–914, Online, August 2021c. Association for Computational Linguistics. URL https://aclanthology.org/2021.findings-acl.79.

81. Christian Stab, Tristan Miller, Benjamin Schiller, Pranav Rai, and Iryna Gurevych. Cross-topic argument mining from heterogeneous sources. In *Proceedings of the 2018 Conference on Empirical Methods in Natural Language Processing*, pages 3664–3674, Brussels, Belgium, October-November 2018. Association for Computational Linguistics. URL https://aclanthology.org/D18-1402.
82. Yonghui Wu, Mike Schuster, Zhifeng Chen, Quoc V Le, Mohammad Norouzi, Wolfgang Macherey, Maxim Krikun, Yuan Cao, Qin Gao, Klaus Macherey, et al. Google's neural machine translation system: Bridging the gap between human and machine translation. *arXiv preprint* arXiv:1609.08144, 2016.
83. Satanjeev Banerjee and Alon Lavie. Meteor: An automatic metric for mt evaluation with improved correlation with human judgments. In *Proceedings of the acl workshop on intrinsic and extrinsic evaluation measures for machine translation and/or summarization*, pages 65–72, 2005.
84. Thibault Sellam, Dipanjan Das, and Ankur Parikh. Bleurt: Learning robust metrics for text generation. In *Proceedings of the 58th Annual Meeting of the Association for Computational Linguistics*, pages 7881–7892, 2020.
85. Kishore Papineni, Salim Roukos, Todd Ward, and Wei-Jing Zhu. Bleu: a method for automatic evaluation of machine translation. In *Proceedings of the 40th annual meeting of the Association for Computational Linguistics*, pages 311–318, 2002b.
86. James N Farr, James J Jenkins, and Donald G Paterson. Simplification of flesch reading ease formula. *Journal of applied psychology*, 35(5):333, 1951.

# Appendix A
# Data Repository

This primarily contains some of the popular data repositories that have been used in this book or are popularly used in the literature. First, we discuss the popular hate speech datasets:-

## A.1 Automated Hate Speech Detection and the Problem of Offensive Language

Davidson et al. [1] built a large dataset of around 24,000 tweets containing hate speech, offensive and normal text. Each tweet was coded by three or more people. only around 5% of the posts were coded as hate speech, out of which 1.3% was unanimously voted as hate speech. Most of the content around 76% was considered as offensive speech.

## A.2 Learning from the Worst (Dynamically Generated Hate Speech Dataset)

This dataset [2] contains a synthetically created dataset by several crowd annotators, where the primary aim of the task was to write a complex hate speech that can make the model predict wrong. Doing this task in four different rounds, the authors collected a dataset of 41255 posts. In each round, they used the dataset collected in the previous rounds to better the classifier. This dataset also collects hate speech from different targets.

## A.3 Large Scale Crowdsourcing and Characterization of Twitter Abusive Behaviour

This is another dataset [3] which contains 80,000 posts from Twitter and has seven different labels—offensive, abusive, hateful speech, aggressive, cyberbullying, spam, and normal. Each tweet was initially coded by 5 annotators. More annotators were employed in order to give the final label if the judgement was not reached in the first round. The dataset contained 18% of the posts as abusive (not normal or spam) and was primarily in English.

## A.4 HateXplain: A Benchmark Dataset for Explainable Hate Speech Detection

This dataset [4] contains 20,000 posts having labels, explanations and targets. The labels were one of hate speech, offensive or normal. The targets comprise various targets with the broader categorization of race, religion, gender, sexual orientation and others. The explanation contains spans of text which justifies the text as hate speech and offensive. The data was collected from the Gab and X. Each post was annotated by 3 annotators.

## A.5 AMI @ EVALITA2020: Automatic Misogyny Identification

This dataset [5] which contains around 7000 posts which have binary labels—misogyny vs non-misogyny. The dataset has 47–50% as abusive language. It is entirely in Italian and the posts were collected from X. There were two tasks in this dataset, one was to detect misogyny or aggressive posts (this was separate) and the second one was to do the same task in an unbiased way. They ran this as a shared task in EVALITA 2020.

## A.6 HatEval: Multilingual Detection of Hate Speech Against Immigrants and Women in X

This dataset [6] contains 19,600 posts in X, 13,000 for English and 6,600 for Spanish. Along the target, 9,091 posts were about immigrants and 10,509 posts were about women. The authors relied on annotators from the figure8 (http://www.figure-eight.com/) platform. This was published as a shared task in SemEval 2019.

## A.7 BanglaAbuseMeme: A Dataset for Bengali Abusive Meme Classification

This dataset [7] consists of 4043 Bengali memes out of which 1,515 have been labeled as abusive and the remaining 2,528 as non-abusive. Next, 1,664 memes are labelled as sarcastic, while the remaining 2,379 are labelled as not sarcastic. Further, 1,171 memes are labelled as vulgar, while 2,872 memes are labelled as not vulgar. Finally, 592 memes are labelled as having a positive sentiment, 1,414 memes as neutral, and 2,037 memes as having a negative sentiment. We achieved an inter-annotator agreement of 0.799, 0.801, 0.67, and 0.72 for the abusive, vulgar, sarcasm, and sentiment labelling tasks, respectively, using the Fleiss' $\kappa$ score. The dataset has been collected from Google, Bing, Facebook, and Instagram.

## A.8 HateMM: A Multi-Modal Dataset for Hate Video Classification

This work [7] relies on BitChute for the collection of this dataset. It serves as a video hosting and sharing platform similar to YouTube and is quite popular among far-right users. The authors curate one of the largest known datasets of hateful videos consisting of 1083 videos spanning $\tilde{4}3$ h and $\tilde{1}44$K frames. Each video was annotated as hateful or not, along with the frame spans that justify the labelling decision.

Next, we note down different **counterspeech** datasets. These are primarily datasets having hate speech-counterspeech pairs or counterspeech/non-counterspeech posts. The former can be used for counterspeech generation while the latter can be used for counterspeech detection or quality evaluation.

## A.9 Thou Shalt Not Hate: Countering Online Hate Speech

This dataset [8] contains a list of counterspeech vs non-counterspeech posts. The annotators further classify the counterspeech into 8 different types. In total, they had around 12k posts from YouTube out of which around 6898 were counter speech. The posts were in response to different hateful videos on YouTube.

## A.10 CONAN—COunter NArratives Through Nichesourcing: A Multilingual Dataset of Responses to Fight Online Hate Speech

This dataset [9] contains a multilingual counterspeech dataset having three languages—English, French and Italian. Overall they contain 15024 hate speech-counterspeech pairs with 6654, 5157, and 3213 in English, French and Italian language respectively. The hate speech and the counterspeech were written by a mix of experts and non-expert users.

## A.11 A Benchmark Dataset for Learning to Intervene in Online Hate Speech

This was one of the first datasets [10] proposed to generate counterspeech. They employed several crowd workers to and asked them to write counterspeech. Overall they collected datasets from two sources Gab and Reddit. The total size of the Gab data was 33,776 out of which 0.43 were abusive. The total size of the Reddit data was 22,324 out of which 0.23 were abusive. The annotators were asked to provide intervention suggestions to these posts.

## A.12 Human-in-the-Loop for Data Collection: A Multi-target Counter-Narrative Dataset to Fight Online Hate Speech

This dataset [11] contains multi-target counter-narrative (speech) generation data. It contains 5003 hate speech counterspeech pairs which have the hate speech based on different protective attributes—race, religion, country of origin, sexual orientation, disability, gender etc. Both the hate speech and the counter speech were mostly written by experts.

More datasets can be found at the following website.[1]

## References

1. Thomas Davidson, Dana Warmsley, Michael Macy, and Ingmar Weber. Automated hate speech detection and the problem of offensive language. In *Proceedings of the international AAAI conference on web and social media*, volume 11, pages 512–515, 2017.
2. Bertie Vidgen, Tristan Thrush, Zeerak Waseem, and Douwe Kiela. Learning from the worst: Dynamically generated datasets to improve online hate detection. In Chengqing Zong, Fei Xia, Wenjie Li, and Roberto Navigli, editors, *Proceedings of the 59th Annual Meeting of the Association for Computational Linguistics and the 11th International Joint Conference on Natural Language Processing (Volume 1: Long Papers)*, pages 1667–1682, Online, August 2021. Association for Computational Linguistics. URL https://aclanthology.org/2021.acl-long.132.
3. Antigoni Founta, Constantinos Djouvas, Despoina Chatzakou, Ilias Leontiadis, Jeremy Blackburn, Gianluca Stringhini, Athena Vakali, Michael Sirivianos, and Nicolas Kourtellis. Large scale crowdsourcing and characterization of twitter abusive behavior. In *Proceedings of the international AAAI conference on web and social media*, volume 12, 2018.
4. Binny Mathew, Punyajoy Saha, Seid Muhie Yimam, Chris Biemann, Pawan Goyal, and Animesh Mukherjee. Hatexplain: A benchmark dataset for explainable hate speech detection. *arXiv preprint* arXiv:2012.10289, 2020b.
5. Elisabetta Fersini, Debora Nozza, Paolo Rosso, et al. Ami@ evalita2020: Automatic misogyny identification. In *Proceedings of the 7th evaluation campaign of Natural Language Processing and Speech tools for Italian (EVALITA 2020)*. (seleziona...), 2020.
6. Valerio Basile, Cristina Bosco, Elisabetta Fersini, Debora Nozza, Viviana Patti, Francisco Manuel Rangel Pardo, Paolo Rosso, and Manuela Sanguinetti. SemEval-2019 task 5: Multilingual detec-

---

[1] https://hatespeechdata.com/.

tion of hate speech against immigrants and women in Twitter. In Jonathan May, Ekaterina Shutova, Aurelie Herbelot, Xiaodan Zhu, Marianna Apidianaki, and Saif M. Mohammad, editors, *Proceedings of the 13th International Workshop on Semantic Evaluation*, pages 54–63, Minneapolis, Minnesota, USA, June 2019b. Association for Computational Linguistics. URL https://aclanthology.org/S19-2007.
7. Mithun Das and Animesh Mukherjee. Banglaabusememe: A dataset for bengali abusive meme classification. In *Proceedings of the 2023 Conference on Empirical Methods in Natural Language Processing*, pages 15498–15512, 2023a.
8. Binny Mathew, Punyajoy Saha, Hardik Tharad, Subham Rajgaria, Prajwal Singhania, Suman Kalyan Maity, Pawan Goyal, and Animesh Mukherjee. Thou shalt not hate: Countering online hate speech. In *Proceedings of the international AAAI conference on web and social media*, volume 13, pages 369–380, 2019b.
9. Yi-Ling Chung, Elizaveta Kuzmenko, Serra Sinem Tekiroglu, and Marco Guerini. CONAN - COunter NArratives through nichesourcing: a multilingual dataset of responses to fight online hate speech. In *Proceedings of the 57th Annual Meeting of the Association for Computational Linguistics*, pages 2819–2829, Florence, Italy, July 2019. Association for Computational Linguistics. URL https://www.aclweb.org/anthology/P19-1271.
10. Jing Qian, Anna Bethke, Yinyin Liu, Elizabeth Belding, and William Yang Wang. A benchmark dataset for learning to intervene in online hate speech. In *Proceedings of (EMNLP-IJCNLP)*, pages 4755–4764, November 2019a.
11. Margherita Fanton, Helena Bonaldi, Serra Sinem Tekiroğlu, and Marco Guerini. Human-in-the-loop for data collection: a multi-target counter narrative dataset to fight online hate speech. In *Proceedings of the 59th Annual Meeting of the Association for Computational Linguistics and the 11th International Joint Conference on Natural Language Processing (Volume 1: Long Papers)*, pages 3226–3240, Online, August 2021. Association for Computational Linguistics. URL https://aclanthology.org/2021.acl-long.250.

The manufacturer's authorised representative in the EU is Springer Nature Customer Service Centre GmbH, Europaplatz 3, 69115 Heidelberg, Germany. If you have any concerns regarding our products, please contact ProductSafety@springernature.com

Printed and bound by CPI Group (UK) Ltd, Croydon, CR0 4YY

26/03/2026

02078939-0020